M140

Introducing statistics

Book 2

Regression and surveys

This publication forms part of an Open University module. Details of this and other Open University modules can be obtained from the Student Registration and Enquiry Service, The Open University, PO Box 197, Milton Keynes MK7 6BJ, United Kingdom (tel. +44 (0)845 300 6090; email general-enquiries@open.ac.uk).

Alternatively, you may visit the Open University website at www.open.ac.uk where you can learn more about the wide range of modules and packs offered at all levels by The Open University.

To purchase a selection of Open University materials visit www.ouw.co.uk, or contact Open University Worldwide, Walton Hall, Milton Keynes MK7 6AA, United Kingdom for a brochure (tel. +44 (0)1908 858779; fax +44 (0)1908 858787; email ouw-customer-services@open.ac.uk).

The Open University, Walton Hall, Milton Keynes, MK7 6AA.

First published 2013.

Edited, designed and typeset by The Open University, using the Open University TEX System.

Printed in the United Kingdom by Charlesworth Press, Wakefield.

ISBN 978 1 7800 7457 3

1.1

Contents

Unit 4

Surveys

Introduction

Units 1–3 have been largely concerned with stage 3 of the modelling diagram (shown in Figure 1), the analysis of the data.

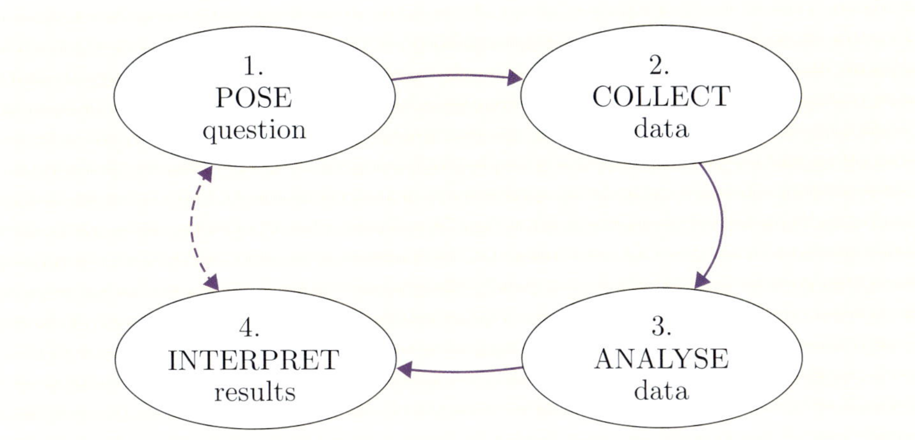

Figure 1 Modelling diagram

This unit concentrates on stage 2, collecting the data. You should by now realise the importance of collecting data that

- can be analysed
- enable you to answer the question under investigation.

Perhaps the most frequent contact that you have with data collection in your everyday life is when you fill in forms or answer questionnaires providing information about yourself, your home, your job, your car or (almost certainly) your OU studies! These can be online or paper and may be for market research companies, government departments or your employers.

Often you are asked to supply the information because you have been selected as one of a relatively small number of people being surveyed, i.e. a **sample**. In other cases, such as the ten-yearly Census in the UK (logos shown in Figure 2), you are part of a large exercise designed to collect information from as many people in the country as it is possible to reach. We shall use the word **census** for any such complete coverage of a population and the word **survey** when a sample is selected from the population.

Figure 2 The 2011 Census logos for England, Wales and Northern Ireland, and for Scotland

You may well have wondered, when you are selected to answer questions in a survey, how the answers you give (about your preferences in toothpaste, or the number of children you have) will affect decisions made by whoever commissioned the survey. You may also have considered the question: if your next-door neighbour had been selected instead of you, how much difference would this have made to any decision based on the survey's results? The results of surveys of one kind or another – opinion polls, advertisers' claims – are often in the news; but do they mean anything useful?

Which was more impressive, the Tower of Suurhusen or the Tower of Pisa?
If undecided, which way did you lean?

Turning these questions about surveys round and looking at them from the statistician's viewpoint leads to the following question.

> *Is it possible to gain useful information about a large population (such as all the people in the UK, or all the employees of a large firm) by collecting data about only a relatively small number (i.e. a sample) of them?*

The answer, which will be explained in more detail in this unit, is *yes*, provided that the people to be questioned are selected in the correct way.

The **population** need not be a population of people; it could consist of schools, firms, villages, fish, light bulbs, etc. Similar questions can be asked about these populations. For example:

> *Is it possible to gain useful information about how long light bulbs will last by testing a relatively small number of them?*

The answer is again yes, provided that the particular items measured or tested are selected in the correct way. Here, though, we shall concentrate on surveys of people.

Section 1 of this unit describes the basic principles of how to select the people to be questioned and introduces a method called *random selection*, or *random sampling*. Section 2 examines the effects of simple random sampling and introduces a modification of this method, called *systematic random sampling*, which is of great practical importance. Section 3 looks more closely at the relationship between samples of the population and the population as a whole. This leads to the idea of a *sampling distribution*, which forms the theoretical basis of methods given in later units for deriving information about the whole of a large population from facts about a sample taken from it. Section 4 contains an introduction to some further aspects of survey planning. Finally, Section 5 directs you to the Computer Book. You are also guided to the Computer Book at the end of Section 3 as you can choose to work through it from this point if you like.

1 Surveys and sampling

Throughout the previous units, emphasis has been laid on the importance of collecting data that are both relevant to the investigation in hand and reliable. You have also encountered several published sources of data. Now, many of these published sources were based on data that had been collected in surveys. Here is a list of those surveys that have been referred to, with a brief description of them.

1. The survey of prices, carried out each month by a market research company on behalf of the Office for National Statistics; this provides over 100 000 prices used in calculating the Retail Prices Index (**RPI**) and the Consumer Prices Index (**CPI**). (See Section 5 of Unit 2.)
2. The Living Costs and Food Survey (**LCF**), which collects information on the spending pattern of 5000 households. (See Section 5 of Unit 2.)
3. The Annual Survey of Hours and Earnings (**ASHE**) which, each year, collects information on the earnings of about 180 000 people. (See Section 1 of Unit 3.)
4. The Monthly Wages and Salaries survey (**MWSS**) which, each month, collects information about the weekly wages of all employees in about 9000 businesses for use in calculating the Average Weekly Earnings (**AWE**). (See Section 5 of Unit 3.)

All these sources of data have one thing in common: they do *not* collect information about *every* individual member of the population involved (i.e. they are *surveys*, not *censuses*). The whole population of interest is known as the **target population**. Each of these surveys claims to provide reliable information about the *whole* of its target population.

1. For the survey of retail and consumer prices, the exact size of the whole target population is difficult to assess but it is certainly much larger than the 100 000+ prices collected in the survey.
2. The target population of the LCF is all households in the UK. There are about 23 000 000 (23 million) of these.
3. The target population of the ASHE is all employees in the UK. There are about 29 000 000 (29 million) of these.
4. Since the AWE aims to give an overall measure of changes in the wages and salaries of all employees in the UK, the target population is all businesses in the UK. Altogether, there are about 4 800 000 (4.8 million) businesses in the UK. Although businesses employing fewer than 20 people are not sampled, the survey covers approximately half of those in employment in the UK.

The basis for using a survey instead of a census is that, provided the sample is chosen carefully from the target population, the results of the survey can be used to infer the characteristics of the whole target population. We shall see later how this can be done, but first let us consider some of the advantages.

1.1 Why do a survey?

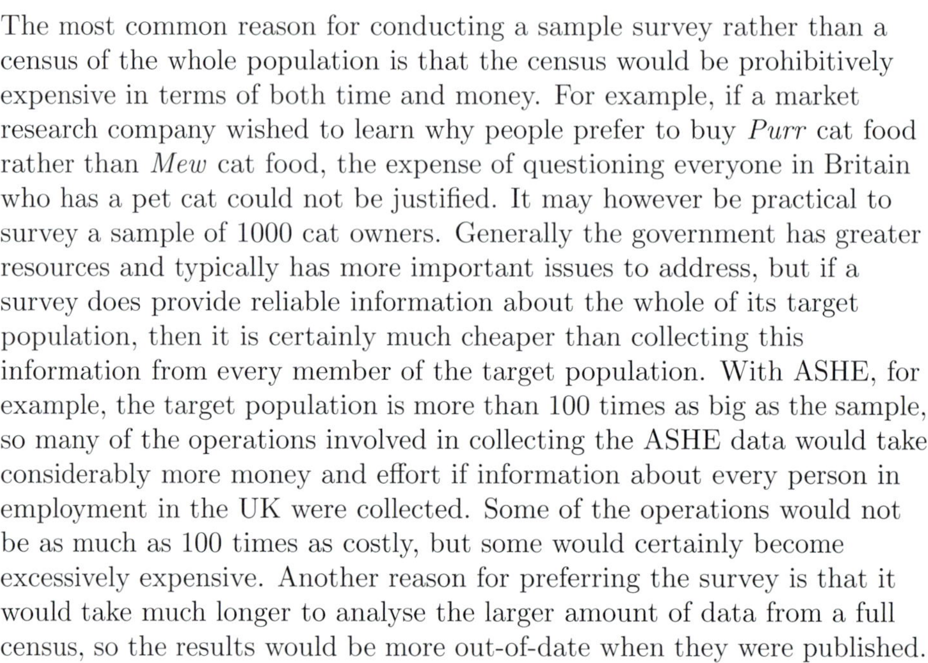

The most common reason for conducting a sample survey rather than a census of the whole population is that the census would be prohibitively expensive in terms of both time and money. For example, if a market research company wished to learn why people prefer to buy *Purr* cat food rather than *Mew* cat food, the expense of questioning everyone in Britain who has a pet cat could not be justified. It may however be practical to survey a sample of 1000 cat owners. Generally the government has greater resources and typically has more important issues to address, but if a survey does provide reliable information about the whole of its target population, then it is certainly much cheaper than collecting this information from every member of the target population. With ASHE, for example, the target population is more than 100 times as big as the sample, so many of the operations involved in collecting the ASHE data would take considerably more money and effort if information about every person in employment in the UK were collected. Some of the operations would not be as much as 100 times as costly, but some would certainly become excessively expensive. Another reason for preferring the survey is that it would take much longer to analyse the larger amount of data from a full census, so the results would be more out-of-date when they were published.

It is certainly true that since only part of the population is included in the sample, the accuracy of the results is threatened, as the characteristics of a sample are very unlikely to be *exactly* those of the whole target population. However, if a suitable method of selection is used in choosing the sample, it is possible to be fairly precise about how large a discrepancy is likely to occur between certain characteristics of the sample and the corresponding characteristics of the target population. The sampling method can then be planned in such a way that the results of the survey are accurate enough for the purpose for which they are needed. Also, in a survey, more care and attention can be given at an individual level than is feasible in a census. This should improve the quality of the data that are gathered, and this will partly offset the uncertainty that arises from sampling.

1.2 Random sampling

In choosing the sample of people to be questioned in a survey, it is important that a suitable method of selection is used. If the statisticians working on the ASHE chose their sample of employees by asking every business in the country how much the managing director earned, for example, then the data collected would not be a very useful measure of the distribution of earnings in the country! A useful sample must be spread evenly over the target population. However, the ASHE statisticians would still not get very accurate information about earnings in the country as a whole by investigating the earnings of a sample of, say, just five people, however carefully they were selected. A useful sample must also be large enough – but how large is large enough? How should a sample be chosen to obtain accurate information about a large population, within constrained budgets?

We require a method of choosing a sample from the target population that is no larger than necessary, because, in general, the smaller the sample, the cheaper the collection of the data. On the other hand, the information collected from the sample must enable us to obtain sufficiently accurate information about the target population; and this means that we cannot choose very small samples. The size of the sample used in a survey has to be a compromise between these two criteria, which can be summarised as **economy** and **accuracy**. Resolving the conflict between these criteria is the aim of a good method of choosing a sample.

Other factors that affect the cost of a survey will be considered in Section 4.

The process of carrying out a survey can be briefly described as follows. You start with a target population and from it you select a sample. You then collect data about this sample. From these data, you want to be able to obtain information about the target population. This process is called *inferring back from the sample to the population.* So you want to choose a sample with properties similar to those of the target population.

The ASHE uses the sample of all people whose National Insurance number ends in a particular pair of digits. This is a good method of choosing a sample for the following reason: there is no relationship between people's National Insurance numbers and their earnings, and this implies that the distribution of the earnings of people in this sample is very likely to be similar to the distribution of the earnings of the whole target population. A slightly more precise way of expressing this property of an ideal sample is to say that *a pattern in the sample implies a similar pattern in the target population.* Such a sample is called a **representative sample**.

The UK does not have a system of personal identity cards, and not everyone has a passport. National Insurance numbers and National Health Service numbers are the only two systems that provide almost every adult in the UK with a code number.

No method of selecting the members of a sample can be guaranteed *always* to produce a representative sample (unless we select every member of the target population!) but one way of getting close to this ideal is to use a method called *random sampling.* This method will be illustrated by using a very small target population consisting of a fictional household, which contains only four members:

Jim Susan Linda Matthew.

Suppose, for the sake of illustration, that we want to investigate the miserliness of this household by asking a sample of individuals from it how mean they are, but that there is only enough money in our survey budget to draw a sample of two people from the household. (Times are hard.)

In this simple situation, we can write down a list of all the possible samples of two different people that we could choose. There are six of them. They are:

1	Jim	Susan
2	Jim	Linda
3	Jim	Matthew
4	Susan	Linda
5	Susan	Matthew
6	Linda	Matthew

As the name 'random sampling' suggests, we let chance choose our sample for us. We shall introduce chance into our method of selection by throwing a die.

'Die' is the singular of 'dice'. A die is therefore one of those little cubes with dots on its faces (Figure 3). Some people use 'dice' as the singular, but statisticians tend to prefer the former.

Figure 3 A pair of dice

First, we must label the six possible samples from the household with the numbers on the six faces of the die: 1, 2, 3, 4, 5, 6. It does not matter which sample gets each label but we shall use the labelling in the list above. Then we can relate the throwing of any one of these numbers on the die to the selection of a particular sample. If we throw a 3, then we select Jim and Matthew.

So long as we do not cheat when throwing the die, and so long as the die is not 'loaded' in some way that makes some numbers more likely to come up than others, this method of choosing a sample is an example of **random sampling**, and the sample chosen is a **random sample**. Such a method is also called **random selection**, and we say that the members of the sample are selected, or chosen, at random – or that they are randomly chosen. The characteristic of a random sample is that every possible sample has the same chance of being selected.

Figure 4 A 20-sided Roman gaming die from the 2nd Century AD

This method of random sampling could, in principle at any rate, be extended to larger samples from larger target populations by using a fair (i.e. not 'loaded') die with more than six faces. For instance, there are 20 different samples of size three that could be drawn from a household with six members, and we could choose one of these samples by listing them all, numbering them from 1 to 20, and rolling a die with 20 faces (such as that shown in Figure 4).

This might just about be feasible, but things quickly get out of hand with populations and samples of the sort of size that are needed in practice. For instance, suppose you wanted to choose a sample of 100 students from an OU module that has 1000 students in all. The number of possible samples is about 6×10^{139}, and it would clearly be impossible either to write out all the possible samples in a list or to construct a die with 6×10^{139} faces to choose one of them at random. Therefore, we have to develop a slightly different way of choosing our sample of two members of the fictional household out of the population of four. This new way will be much easier to extend to larger samples from larger populations.

The number 6×10^{139} would be written down as a 6 followed by 139 zeros.

What we shall do is to choose the individual people to go into our sample one at a time. Look again at the list of all possible samples.

Jim Susan
Jim Linda
Jim Matthew
Susan Linda
Susan Matthew
Linda Matthew

Each individual appears in the same number (three) of the six possible samples. Therefore, all of the four household members are equally likely to appear in any particular sample that we happen to choose. Let us label the household members, rather than the samples, with numbers:

1 Jim 2 Susan 3 Linda 4 Matthew.

To select the first member of our sample, we throw the die and record the number thrown. Then we select the person who is labelled by this number. (We could use a four-faced die for this if we had one, or we could just use an ordinary six-faced die and ignore any throw which resulted in a 5 or a 6.) To select the second member of our sample, we repeat the above process. However, if the die shows the same number as the first selection, we throw again, because we do not want to include the same person in our sample more than once.

In some circumstances, it *is* appropriate to allow samples in which the same individual can appear more than once, though these types of situations are not considered in this unit.

If we require a sample of size two and the numbers thrown were 2 and 3, then Susan and Linda would be selected. If, however, the numbers thrown were 1 and 1, we would ignore the second 1 and throw again. If we obtained the number 4 on the next throw the sample would be Jim and Matthew. Choosing the sample members one at a time like this still has the property that any of the possible samples is just as likely to be chosen as any other, so that conceptually it is no different from the first method we described. It is much more practical to use this one-at-a-time method for larger samples and populations.

We could choose a sample of three people from a household of size six by numbering the individuals in the household from one to six and throwing a six-sided die at least three times. (More than three throws might be needed to avoid repetitions.) Even for the problem of drawing a sample of 100 students from a population of 1000, the one-at-a-time approach would save having to write out all 6×10^{139} possible samples in a list: we would just have to write out a list of all 1000 students, number them from one to 1000 and start rolling a 1000-faced die. For a target population of 1 000 000 people we should need a die with 1 000 000 faces!

It may seem impossible to do anything like this! In practice, statisticians use computer programs to generate random numbers which can act in this manner. We shall now see how to use random numbers in this way.

You will learn how to use Minitab to generate random numbers in the final section of this unit.

The following random numbers are taken from a set that were generated using Minitab:

9 8 0 6 7 7 4 6 1 6 ....

They can be written as pairs of digits,

98 06 77 46 16 ...,

and are then exactly equivalent to the results of throwing an imaginary fair die with 100 sides labelled 00, 01, 02, ..., up to 99. If you had a target population of size 100, you would probably find it simplest to label the first member 01, the second 02 and so on, with the 99th member labelled 99. Then the 100th member would use the label 00. Then you could use the throws of the imaginary die to select a random sample. As with the real die, if a pair of digits that you have already used in the sample turns up again, you just ignore it and go on to the next pair.

We can also use pairs of digits for target populations of size less than 100, as will be described in Section 2.

So the pairs of digits at the start of the first row in the list above would select those members of the population labelled 98, 06, 77, 46 and 16. These members therefore form a random sample of size five.

If more than one sample is required from the same target population, then you should not start from the same place in your list of random numbers every time, because this would lead to the selection of the same members of the population in every sample. It is important to start at a *different* point in the list for each sample. The starting point should ideally be selected randomly (using a die or some other procedure). However, to aid explanation, you will usually be told where to start in each case.

Activity 1 *Random sample from population of 100*

Choose a random sample of size 12 from the population of 100 individuals labelled 00 to 99, using the method described above. A table of random numbers, generated using a computer, is provided as an appendix to this unit. Use successive pairs from row **79** of the random number table, beginning with the first pair in the row, i.e. 52.

You may have found it a little awkward in the last activity to check for repetitions in the sample. In relatively small samples from larger populations than this, repetitions are very rare occurrences in practice.

For a target population of size 1 000 000, we need to use the following labels

00 00 00 00 00 01 00 00 02 ... up to 99 99 97 99 99 98 99 99 99.

Again, the population would probably be labelled 00 00 01, 00 00 02, 00 00 03, . . . , up to 99 99 99, 100 00 00, and we should use the random number 00 00 00 for the last member. Then, for the throws of an imaginary die with 1 000 000 sides, we use groups of six digits in the random number table. If we start with the row designated **20**, say, then the first three labels selected will be

59 70 46   36 67 19   12 59 39.

Lottery draws

Major lotteries, such as the UK National Lottery, use special machines to draw the random winning numbers. The draws are open (they are often televised), and the purpose of the machines is partly to put on a spectacle but also to make it transparent that the lottery is fair and the numbers are drawn truly at random. The latter is important as a randomly drawn set of numbers will sometimes look very odd. For example, the six numbers drawn in the UK National Lottery on 11 October 2008 (excluding the 'bonus ball') were all in the twenties – 20, 21, 23, 24, 27 and 28 – despite being a random selection from the numbers 1 to 49.

A UK National Lottery machine

Activity 2 *Random sample from population of 1 000 000*

Choose a random sample of size ten from the target population of size 1 000 000 using labels as described above. Use rows **15** and **16** of the table in the same way as we used row **20** above.

1.3 Properties of simple random sampling

You have learned how to find a random sample of the target population (and been told why it is called a random sample). This process is usually called **simple random sampling** (and the samples chosen are called **simple random samples**) to distinguish it from other random methods, some of which will be described later. The very important *random* nature of the procedure can be more precisely expressed as follows.

In Unit 6, we shall be able to express these properties even more precisely because there we shall encounter probability. This is a measure of chance and it gives us a language for describing random processes.

Simple random sampling

This is a method of selecting a sample in which the possible samples of a given size, n, consist of all possible selections of n different individuals from the population. The sample to be used is chosen in such a way that every possible sample is equally likely to be selected.

One way of doing this is to choose the sample members one at a time in such a way that:

- At each selection, every member of the target population is equally likely to be selected.
- The selection of a particular member of the target population has no effect on the other selections, beyond the requirement that the same individual cannot appear more than once in the sample.

It may seem paradoxical to you that we should be recommending a method of obtaining a representative sample in which chance plays such an important role. One analogy that might help you to see why simple random sampling is sensible is the following.

Figure 5 A hand of cards

The process of shuffling a pack of cards well and then dealing a hand is essentially a method of choosing a hand of cards (such as that in Figure 5) by simple random sampling from the pack. If you have played any card game, you will probably be aware that most hands of cards contain a fairly even distribution of suits, and contain a few court cards but not a great many of them. Therefore, they have properties that match the properties of the whole pack, which has an even distribution of suits and just under

25% of the pack is court cards. To put it another way, if you actually wrote down a list of all possible hands of cards, some of them would be unrepresentative in terms of suit distribution or the number of court cards, but most would be representative. Therefore, when one of the possible hands is chosen or dealt at random, it is more likely to be representative than it is to be peculiar.

The characteristics of the collection of all possible samples is dealt with more precisely in Section 3.

In the next section we shall look critically at simple random sampling, and see that it is certainly no exception to the statement made earlier: that no method is guaranteed always to produce a representative sample, i.e. a sample from which we can make completely accurate inferences about the population. (Hands of cards consisting entirely of one suit do turn up!) However, randomness is an essential feature of most good methods of choosing a sample.

It is not always necessary, or possible, to use random numbers to choose a random sample. For example, suppose that you wanted to choose a random sample of size ten from a population of 100 fish in a tank. It would probably be very difficult to label each individual fish, and it would be impossible if you wanted to choose a sample of fish from the North Sea.

Figure 6 Netting fish

It would therefore be impossible to use random numbers to choose a sample. Simply selecting ten fish from some caught in a net (Figure 6) is, for many purposes, as good a method as any of choosing this random sample. Unless, for example, you want to measure their size, or how difficult they are to net!

Much of this section has been concerned with general methods. You have seen that a well-chosen sample is an economic and accurate method of collecting data about a population, and that simple random sampling is a good method of choosing a sample. You have seen how to use random numbers to choose a simple random sample from a population with numerical labels. In contrast, the next section will be more specific and more practical. We shall concentrate on a particular target population and choose some random samples from it.

Exercises on Section 1

Exercise 1 *Random sample from population of 1000*

In this exercise we have a new target population whose size is 1000. Use the random number table in the appendix to choose a random sample of size seven from this population.

Exercise 2 *Random sample from population of 100*

The population in this exercise is of size 100, labelled 00 to 99.

(a) Choose a random sample of size nine using pairs from row **25**. Start at the third pair, which is 26, and work to the right.

(b) Choose a random sample of size 17 using pairs starting at the beginning of row **26**. Move along row **26** to the right-hand end and then go to the next row, designated **27**.

2 Random samples

Throughout this section, we shall assume that, just as in Units 2 and 3, we are interested in investigating whether people have been getting better or worse off. To pursue this investigation, we might carry out a survey in which several related, and relevant, questions on this subject are put to a sample of individuals. The questions might be concerned with changes in their income and expenditure, as well as their subjective feelings about their economic well-being.

Our target population will be those people who work in the mythical Sampling Department in a large organisation. These 86 people are listed in Table 1 in alphabetical order of surname. This list is based on a staff list from a real organisation; the names and other details have been changed to preserve confidentiality.

Each person has been given a label. We have also recorded their gender and occupational group. The information in these last two columns will not be used immediately; it will become relevant later, because a person's gender and occupation may have a bearing on how well off he/she is. For choosing a random sample, we need the second column together with some random numbers. We will use the table of random numbers given in the appendix to this unit.

Table 1 Sampling Department staff list (in alphabetical order)

Name	Label	Gender	Occupation*
Alicante-Node, Alphonso	01	M	M
Andrews, Jean	02	F	P
Archer, Simon	03	M	M
Baines, Tom	04	M	P
Baker, Fred	05	M	P
Bates, Sheila	06	F	S
Baxter, John	07	M	P
Best, John	08	M	P
Bidford, David	09	M	P
Bond, Mick	10	M	P
Bramley, Max	11	M	P
Burroughs, Sean	12	M	P
Cameron, Lynne	13	F	P
Carter, Jane	14	F	P
Chapman, Liz	15	F	M
Clark, Rowena	16	F	S
Clarke, Jim	17	M	A
Cluskie, Alex	18	M	P
Cramer, Will	19	M	P
Crofts, Dennis	20	M	P
Crofts, Mary	21	F	A
Crossman, Kim	22	M	S
Daley, Stuart	23	M	P
Damper, Emma	24	F	S
Dev, Mohen	25	M	P
Eisenstein, Bert	26	M	P
Eric, Steve	27	M	P
Estover, Matthew	28	M	P
Fallow, Jim	29	M	P
Flint, Gerald	30	M	P
Foster, Sue	31	F	S
Franks, Abraham	32	M	P
Gowan, Dai	33	M	P
Graham, Bert	34	M	P
Graham, Bill	35	M	P
Grant, Lynne	36	F	P
Gray, Chris	37	M	P
Greenson, Denise	38	F	A
Greenway, Maggie	39	F	P
Hallow, Jean	40	F	A
Hare, Dorothy	41	F	P
Harrison, Sheila	42	F	P
Hewitt, Ray	43	M	P

* P = Professional, A = Administrative, S = Secretarial, M = Manual

Name	Label	Gender	Occupation*
Hopkins, Jane	44	F	A
Howe, Phil	45	M	P
Hutton, Joan	46	F	S
Iron, Donald	47	M	P
James, Patricia	48	F	A
Jolly, Susan	49	F	S
Kapoor, Sashi	50	M	P
Lang, Chris	51	M	P
Light, Phil	52	M	P
Locke, Carol	53	F	S
London, Fred	54	M	P
Lupton, David	55	M	P
McCarthy, Keith	56	M	P
McCraig, Frank	57	M	P
Masterton, Dick	58	M	P
Menton, Christine	59	F	S
Menton, Pete	60	M	P
Munn, Sharon	61	F	P
Neilsen, Rob	62	M	P
Osterley, Rebecca	63	F	S
Patel, Deepak	64	M	P
Pinder, Andrew	65	M	P
Redman, Guy	66	M	P
Redstar, Pamela	67	F	S
Ricardo, Dan	68	M	P
Roberts, Christine	69	F	S
Rowan, George	70	M	P
Sandford, Dave	71	M	P
Shah, Anjali	72	F	S
Singh, Meera	73	F	S
Stratford, Peter	74	M	P
Thompson, Anna	75	F	S
Thompson, Jack	76	M	P
Trumpington, Pat	77	F	S
Truscott, Karen	78	F	S
Turner, Richard	79	M	P
Tyndale, Babs	80	F	S
Watson, Eleanor	81	F	P
Wilton, Larrie	82	F	P
Winston, Chuck	83	M	P
Woodhouse, Paul	84	M	M
Wu, C. C.	85	F	M
Yeo, Tara	86	F	A

* P = Professional, A = Administrative, S = Secretarial, M = Manual

2.1 Choosing some samples

In Subsection 1.2 we described a way of using random numbers to choose a sample from a target population of size 100. A small adaptation of this method will enable you to choose a sample from the target population of size 86. In the department list (Table 1) the members of the target population are labelled 01, 02, 03, ..., and so on, up to 84, 85, 86. You could therefore use pairs of digits to select members of a sample just as you did for the 100 labels in Subsection 1.2 but, trying this method, if you randomly selected 93 as your starting pair of digits you would be unable to select a person with this label. You should simply ignore this pair and go on to the next pair in your list of random numbers.

To use pairs of digits as throws of an 86-sided die: simply ignore any pair of digits that is not one of the 86 labels in the list of the target population.

Example 1 *Random sample from population of 86*

We shall now use row **53** of the table in the appendix to choose a sample of size ten from our target population. We work along the pairs of digits in this row until we have ten labels in the range 01 to 86, ignoring all pairs of digits outside this range.

An X in the 'Label' row means that a pair of digits has been ignored. We would also have to ignore repetitions, but luckily there are none.

Row **53**	93	46	82	67	64	48	91	74	85	94	40	51	30
Label of selected individual	X	46	82	67	64	48	X	74	85	X	40	51	30

Looking for these labels in the department list we find the sample listed in Table 2. This table shows the name and label of the ten people selected for the sample and also their gender and occupation. The last column, which is headed 'Response', is explained below.

Table 2 A sample of ten staff

Name	Label	Gender	Occupation	Response
Hutton, Joan	46	F	S	No
Wilton, Larrie	82	F	P	Yes
Redstar, Pamela	67	F	S	Yes
Patel, Deepak	64	M	P	No
James, Patricia	48	F	A	Yes
Stratford, Peter	74	M	P	No
Wu, C. C.	85	F	M	Yes
Hallow, Jean	40	F	A	Yes
Lang, Chris	51	M	P	No
Flint, Gerald	30	M	P	No

Example 1 is the subject of Screencast 1 for Unit 4 (see the M140 website).

Now that we have selected a random sample of people in the department, we can use it to investigate whether people think they are getting better off. We might start by asking the ten people a straight question, 'Do you feel that you are better off now than you were twelve months ago?' and ask for a straight 'Yes' or 'No' response. Suppose that the answers given to this question are those shown in the last column of Table 2.

We shall discuss the choice of question a little more in Subsection 3.1.

In the sample, there were five 'Yes' responses and five 'No' responses. Can we say that there would be equal numbers of 'Yes' and 'No' responses in the whole population? In other words, how representative is the sample of the target population? Is there anything we can do to check its representativeness? We cannot check whether the responses to the question are representative because we do not know the responses of the whole target population. However, we can use the information in the columns headed 'Gender' and 'Occupation' in Table 1 to check how representative the sample is for these characteristics. If the sample is unrepresentative in terms of gender or occupation, it is less likely to be representative in terms of whether people feel they are getting better off. However, before we can do this check, we must analyse the information contained in these columns. The information contained in Table 1 about the structure of the target population is summarised in Table 3, which lists the number of department staff of each gender and the number in each occupational group; there are eight different gender–occupation categories in all.

Table 3 Department staff analysed by gender and occupation

	Male	Female	Total
Professional	46	10	56
Administrative	1	6	7
Secretarial	1	17	18
Manual	3	2	5
Total	51	35	86

Since the staff list is based on that of a real organisation, it reflects the fact that in many British organisations the gender balance in different occupations remains uneven. Out of 56 people in the professional group, 46 (82%) are male, whereas 17 out of the 18 secretarial staff (94%) are female. The module team chose to use this particular example not because we approve of the status quo on gender balance, but because we want to demonstrate the important role that statistics can play in investigating such issues and monitoring change.

Table 3 can be used to compare the target population with any sample from it and thus to check on whether the sample is *representative* with respect to gender and occupation. To do this, it is usually better to express the number in each category as a percentage of the total: 86. This has been done in Table 4.

Table 4 Percentages of department staff by gender and occupation

	Male	Female	Total
Professional	53.5	11.6	65.1
Administrative	1.2	7.0	8.1
Secretarial	1.2	19.8	20.9
Manual	3.5	2.3	5.8
Total	59.3	40.7	100.0

Note that all the percentages in Table 4 were found by dividing the corresponding entry in Table 3 by 86, multiplying by 100 and then rounding to one decimal place. Therefore, some of the figures in the 'Total' row and column of Table 4 do not correspond exactly to the totals of the rounded values in the table, because of the small inaccuracies introduced by rounding.

Using this information we can now demonstrate that the sample in Example 1 is not very representative. Two facts will suffice.

- The majority of the sample – six out of ten, or 60% – consists of women, compared with only 40% of the population.
- 20% of the sample are in the administrative category, and 50% are in the professional category, compared with the proportions of about 8% and 65%, respectively, in the population.

This sample should thus be described as unrepresentative with respect to gender and occupation. It would not be possible to reproduce all the percentages in Table 4 exactly in a sample of only 10, of course, but you might hope to get rather closer than we did in this sample. The sample was chosen by random sampling but it has turned out to be unrepresentative of the population in terms of gender and occupation. Therefore, if you were able to do a similar comparison for responses to the question about how well off people felt, you might well find that the results from the sample did not agree with those of the population.

As you would expect intuitively, all other things being equal, the larger the sample chosen from the population, the more representative it is likely to be, and the closer the characteristics of the sample will be to those of the population.

Activity 3 *Sampling from the Sampling Department*

Choose a random sample of size 20 from the department list using the random number table provided in the appendix to this unit starting at the beginning of row **2**.

Note the gender and occupation of each individual selected and then comment on the representativeness of the sample with respect to gender and occupation.

2.2 Systematic random sampling

You should now be able to appreciate how time-consuming and tedious it would be to choose even a moderately large sample from a fairly large population using simple random sampling. The sizes of the samples we have chosen so far are trivial compared to the sampling requirements of some official, academic and market research investigations.

In practice, for a real survey the sample would be drawn using a computer. Computers do not find jobs tedious (or enjoyable!). In Section 5 you will learn to use Minitab to draw random samples.

An alternative method, which provides a quicker and easier means of choosing a sample from a list of the target population, is **systematic random sampling**. This method is similar to that used to choose the sample for the ASHE (Annual Survey of Hours and Earnings), which selects one in 100 of the National Insurance numbers (which are themselves issued sequentially). The ASHE does not select these labels randomly but selects all the labels with the same pair of final digits. The only randomness in this procedure comes in choosing which one of the 100 pairs of digits to use. Having made this choice, the selection is completely systematic and can be described as selecting every 100th label in the ordered list of labels.

So, as the National Insurance numbers are just labels, we can use the labels 01 to 86 of our population in Table 1 in a similar way.

Example 2 *Sampling every eighth individual*

Using a similar procedure to that above, select a sample of about one-eighth of our target population using the labelled list as follows.

Step 1 Decide where to start by randomly choosing a label from the first eight labels, 01 to 08. This label is the **random start**. Suppose that it is 04.

Step 2 Select the remaining individuals from the population by systematically selecting every eighth label. The number eight is the **sampling interval**.

This gives the following 11 labels.

04 12 20 28 36 44 52 60 68 76 84

So the sample with sampling interval eight and random start 04 is as shown in Table 5.

Table 5 A sample of every eighth individual

Name	Label	Gender	Occupation
Baines, Tom	04	M	P
Burroughs, Sean	12	M	P
Crofts, Dennis	20	M	P
Estover, Matthew	28	M	P
Grant, Lynne	36	F	P
Hopkins, Jane	44	F	A
Light, Phil	52	M	P
Menton, Pete	60	M	P
Ricardo, Dan	68	M	P
Thompson, Jack	76	M	P
Woodhouse, Paul	84	M	M

Example 2 is the subject of Screencast 2 for Unit 4 (see the M140 website).

In the sample selected in Example 2 there are nine professionals, one administrator, one manual worker and no secretarial staff. Also, there are nine men and only two women, compared to a ratio in the whole population of six to four. Overall, the sample is not very representative of the whole target population.

This shows that a systematic random sample *need* not be any more representative than a simple random sample. However, there are two main reasons for using systematic random sampling: one is to save time, and the other is that in certain special circumstances (which we shall come to later) systematic sampling *does* tend to produce more representative samples.

This method does not always give samples of exactly the same size. This is illustrated in the following example.

Example 3 *A second systematic sample*

Suppose the random start is 07 and we select every eighth label (i.e. we use the same sampling interval eight). Then we get only these ten labels:

07 15 23 31 39 47 55 63 71 79.

In practice, these discrepancies in size hardly ever matter, as the sample size will only vary by one, and typical sample sizes in actual samples are usually several thousand.

This second sample, with sampling interval eight and random start 07, is shown in Table 6.

Table 6 A second systematic sample

Name	Label	Gender	Occupation
Baxter, John	07	M	P
Chapman, Liz	15	F	M
Daley, Stuart	23	M	P
Foster, Sue	31	F	S
Greenway, Maggie	39	F	P
Iron, Donald	47	M	P
Lupton, David	55	M	P
Osterley, Rebecca	63	F	S
Sandford, Dave	71	M	P
Turner, Richard	79	M	P

In this sample there are seven professionals, one manual worker, two secretarial staff and no administrators. The ratio of men to women is almost exactly that of the whole population. So this happens to be a more representative sample than the previous ones as regards gender and occupation.

Activity 4 *A systematic sample of one-seventeenth*

Select a systematic random sample of about one-seventeenth of the department. To find the random start, take the first pair of digits in the range 01 to 17 from row **3** of the random number table in the appendix to this unit. Analyse the sample with respect to gender and occupation, and comment on how representative it is in these respects.

Activity 5 *A systematic sample of one-quarter*

Choose a systematic random sample of about a quarter of the department. This time, take the first digit in row **29** in the range 1 to 4 as your random start. Analyse the sample with respect to gender and occupation and comment on how representative it is in these respects.

From the last two activities, and the examples of simple random sampling in Subsection 2.1, you should now be able to appreciate that systematic random sampling is much quicker to do 'by hand' than simple random sampling, but that it does not necessarily provide samples which are more representative of the target population.

In some circumstances systematic random sampling will do no better and no worse, on average, than simple random sampling in producing representative samples. However, in other circumstances it might do much worse: for example, suppose that you have a list of people in which each consecutive pair are a married couple with the husband always appearing first and the wife second. If you take a systematic random sample from such a list and the sampling interval is an even number, then the sample will consist entirely of men or entirely of women, depending on whether the random start is an odd or an even number. This shows that care is needed in the use of systematic random sampling: it is hazardous whenever the list of the population contains such regularities. A case as extreme as this could easily be recognised, but if the regularity is less distinct, and hence not noticed, then the problem is more serious.

There are circumstances, though, in which systematic sampling is likely to do *better* than simple random sampling. Suppose that the department list in Table 1 had been ordered by occupation and gender instead of simply being in alphabetical order of names. That is, suppose that all the female professionals were listed first, followed by all the male professionals, then all the female administrators, then all the male administrators and so on. Imagine drawing a systematic sample of a quarter of the department from a list in that order. The sample would inevitably include about a quarter of the female professionals, a quarter of the male professionals, a quarter of the female administrators – in fact, about a quarter of each gender–occupation group. It would therefore be very representative.

This is a kind of *stratified sampling*, a concept you will learn more about in Section 4.

In simple random sampling, all possible samples are equally likely to be chosen. The method tends to work well because most but not all of the possible samples are reasonably representative. In systematic sampling, the number of different samples it is possible to obtain is much smaller. There are only four possible systematic random samples of a quarter of the population in Table 1, because there are only four possible values for the random start. (By contrast, a simple random sample of 21 people from the same population, about a quarter of the population, would be one chosen at random from about 6×10^{19} possible samples.) If the population were listed in gender–occupation order, then all four possible systematic random samples would be representative, so that systematic sampling is bound to do well. However, in a situation like the list of married couples, all possible systematic samples would be unrepresentative, so that systematic sampling is bound to do badly. In many circumstances, though, the population will be listed in some order that has nothing to do with the features of the population it is important to represent; then systematic random sampling is likely to be no more and no less representative than simple random sampling.

To summarise, we have the following properties of systematic random sampling.

Systematic random sampling

Systematic random sampling is easier to carry out than simple random sampling and is very often used for choosing samples from large populations.

- It can produce very unrepresentative samples if the list of the target population is structured in certain ways.
- It produces random samples that are at least as representative as those produced by simple random sampling, provided the target population is listed in a suitable way.
- In certain cases, systematic random samples are considerably more representative than simple random samples.

In this section you have learned how to choose both simple and systematic random samples, using a labelled list of the target population, and you have learned about some of the properties of the two methods.

Exercises on Section 2

Exercise 3 *Selecting more simple and systematic samples*

This exercise is on choosing both simple and systematic random samples. After choosing each of the following samples from the list in Table 1, draw up a table similar to Table 3 (in Subsection 2.1) to analyse the sample by gender and occupation.

(a) Choose a simple random sample of size eight using row **5** starting at the beginning.

(b) Choose a simple random sample of size 12 using row **10** starting at the beginning.

(c) Choose a systematic random sample with sampling interval nine and random start 05.

(d) Choose a systematic random sample with sampling interval ten and random start 08.

3 Patterns in the samples

So far in this unit we have looked at individual samples from a target population and considered whether a sample is representative of its target population. In the last section, some of the samples we drew did seem to be representative of the target population; others did not. In this section we shall take a different view of sampling. We shall consider all the possible samples of a given size that could arise when choosing a sample from a given population. You will see that patterns arise in such collections of all possible samples, and that these patterns provide information about the representativeness of samples. Here, we shall look at samples from one particular population, but similar methods can be used to describe patterns in collections of samples from any population.

3.1 Population values and sample values

In Section 2, the aim of sampling from the population was to investigate whether people were getting better or worse off. (That was why we wanted a sample that was representative in terms of gender and occupation – factors likely to determine how well off someone is.) Here, we shall continue with the same aspect of this investigation: determining people's subjective feelings about changes in their own economic circumstances.

There are several methods of obtaining such information, but, because of its subjective nature, they nearly all involve asking people questions. Therefore, a reasonably good method of obtaining the required information is to question a relatively small sample of the target population. The most straightforward question we could ask on this topic is a question such as the following.

Are you better off than you were twelve months ago?

However, such a blunt question would probably not produce very useful data. There are many reasons for this, but one of the most crucial is that different people will interpret it in different ways. (To test this claim, try asking your friends this question and note the way in which they interpret it.) A better question for our investigation is as follows.

Considering what has happened to your earnings, the way prices have changed and changes in other circumstances, do you feel that you are now better or worse off than you were twelve months ago?

This question still leaves one problem that always occurs when investigating people's subjective feelings. If someone asked you a question like this, you might well reply at length describing your personal circumstances and events during the year. Such responses are hard to analyse, so it is very common to ask the respondent to classify his or her answer into one of a small number of categories.

Rensis Likert (1903–1981)

This is most commonly done through a **Likert scale**, named after Rensis Likert (1903–1981), whose work underlies its popularity.

A Likert scale has a number of ordered categories, and respondents tick one of them to specify their level of agreement or disagreement with a statement. For the above question, the following request could be added.

Please tick the phrase that best describes your feelings.

Much better off	☐
Somewhat better off	☐
About the same	☐
Somewhat worse off	☐
Much worse off	☐

This makes it much easier to compare one person's answer with another's and to summarise people's answers. Analysis of the answers is yet further simplified if each response is expressed as a number from 1 to 5 as follows.

Much better off	5
Somewhat better off	4
About the same	3
Somewhat worse off	2
Much worse off	1

So, the better off a person feels they have become, the higher the number we use to label their response.

There are snags. The simplification obscures the individual details of what people might have said if they had been given the opportunity, and you might still worry about whether one person's 'somewhat better off' is the same as another's.

It is important to realise that the numbers are being used here simply as labels that come in a helpful order. There is no implication that, for instance, 'Somewhat better off' is twice as good as 'Somewhat worse off', just because 4 is twice 2. In fact, the labels for the responses could have been chosen as a, b, c, d, e, rather than 1 to 5.

If we choose a sample of people from the target population and ask them this question, then we shall know what those people's answers are: these are the **sample data**. We shall then wish to infer from these sample data information about how the whole of the target population would have answered this question had we asked them all. More precisely, the **response** to the above question is 1, 2, 3, 4 or 5, and we shall want to infer back from the **sample values** of this response to values of this response for members of the target population as a whole. These values for the whole target population are the **population values** of the response.

3.2 All possible samples

The examples in Section 2 demonstrated that *any* method of choosing the relatively small sample required can produce a sample that is not very representative of the target population. Although the best methods of choosing a sample are designed to produce representative samples as consistently as possible, none of them guarantees to do so without fail. However, the samples we analysed in Section 2 suggest that, for all but the smallest sample sizes, either of the random sampling methods (simple or systematic) is likely to produce a sample that is sufficiently representative to justify inferring back to the population from facts about the sample.

The samples in that section also suggested that if you choose a larger sample, then you are more likely to choose a representative sample. The reason for this is that although the results from an individual, randomly-chosen sample may well have no clear pattern, the results obtained from the collection of *all possible* samples of a fixed size has a very distinctive pattern for all but the smallest sample sizes.

We will examine some of these patterns. To do this, it is necessary to imagine that we know *all* the relevant information about the target population. We can then consider what samples taken from that target population might look like. That is, imagine that a census was carried out in which every individual in the target population was asked the question we are interested in, and that we knew what all the responses were. In the rest of this section we shall take this convenient, though rather unrealistic, omniscient view.

Imagine first that the target population is 1000 individuals whose responses to the question (i.e. the population values of the response) are already known to be as described in Table 7.

Table 7 Population values of the response

Response	Rating	Number
Much worse off	1	300
Somewhat worse off	2	100
About the same	3	200
Somewhat better off	4	300
Much better off	5	100
Total		1000

We are now interested in the responses of *all* the possible samples of a fixed size that could be obtained from this population by simple random sampling. Even for fairly small sample sizes the numbers involved at this stage are quite large. There are 499 500 possible simple random samples of size two, 166 167 000 of size three, 41 417 124 750 of size four, and so on.

With such large numbers of samples to consider, it may seem impossible to deduce anything at all sensible about these collections of all possible samples. This problem is made easier because, very often, our main interest lies in just one, or a few, properties of the sample and the population. Suppose, for instance, that we are particularly interested in the median of the responses for the population, perhaps because we want a measure of location for the population's responses.

Activity 6 *Population median for Likert data*

Find the median of the responses of the population described in Table 7.

The median calculated in Activity 6 is often called the **median of the response over the whole population** (or, more briefly, the **population median response**, because the median of a population is often called the **population median**). It was possible to find the population median response, in the way you have just done, only because we have imagined that we know all the population values of the response. In a practical situation, you would have data from only a sample from the population. You could calculate the median of the responses in the sample, of course, but what would that tell you about the population median response? To answer this question, we need to consider patterns in the medians in the collection of all possible samples.

Many useful methods have been devised to find and describe the patterns in the collection of all possible samples of a fixed size. These methods typically identify properties of interest (such as the property 'median is 3') and then, for each property, calculate the proportion of samples in the collection that have that property. The results from applying one such method will be illustrated in the next subsection, using the target population described in Table 7.

Calculations underlying the method use the rules of probability, which will be introduced in Unit 6.

3.3 Pictures of patterns

Suppose that we choose a very small sample, of size three, from our target population of size 1000. There are 166 167 000 possible samples of size three.

Although not impossible, it would be quite complex to picture the responses of all three individuals in each of these millions of possible samples of size three. It is more straightforward to picture the millions of medians of these sample responses. We can then look for patterns in this batch of medians.

Activity 7 *Median responses in samples of size 3*

Table 8 shows the responses of six typical samples (A to F) of size three from the target population. So, for example, in Sample A the first person who was asked replied 'somewhat better off' and so the result was labelled 4, the second person's response was labelled 1, and the third person's was labelled 2. The median of these three responses is found by rewriting them in numerical order, 1, 2, 4, and then finding the middle value, which is 2.

Write down the median of each of the six batches of sample responses.

Table 8 Responses of the people in six samples of size three

Sample	1st person	2nd person	3rd person
A	4	1	2
B	5	4	1
C	1	4	4
D	1	5	3
E	1	1	1
F	3	5	5

As you have probably realised from this activity, the median of the responses of a sample of size three from this population is either 1, 2, 3, 4 or 5. We shall call such a median a **median response**. It is possible, therefore, to describe the medians of the responses of *all* the 166 167 000 samples of size three by stating how many of them are 1, how many are 2 and how many are 3, 4 and 5. These numbers can be calculated using the rules of probability, and their approximate values are given in Table 9 (where, for example, '359 hundred thousand' means 35 900 000).

Table 9 Median responses of all samples of size three

Median response	1	2	3	4	5
Approximate number of samples (hundred thousands)	359	226	492	539	46

Proportions, such as those used in Table 10, are a very common way of describing such large collections of numbers.

In Table 10 these numbers are expressed as *proportions* of the total number (166 167 000) of samples of size three. This will enable us to look at the pattern, if any, in these sample median responses and to compare the pattern in these medians with the patterns obtained in the same way from samples of other sizes.

Table 10 Median responses of all samples of size three

Median response	1	2	3	4	5
Approximate proportion of samples	0.216	0.136	0.296	0.324	0.028

(These proportions are obtained by dividing the entries in Table 9 by 166 167 000.)

We have displayed these proportions graphically in Figure 7(a), which is a picture of a **sampling distribution**. It is the **distribution of the median response of the sample**; this is often shortened to the

distribution of the sample median (because the median of a sample is often called the **sample median**).

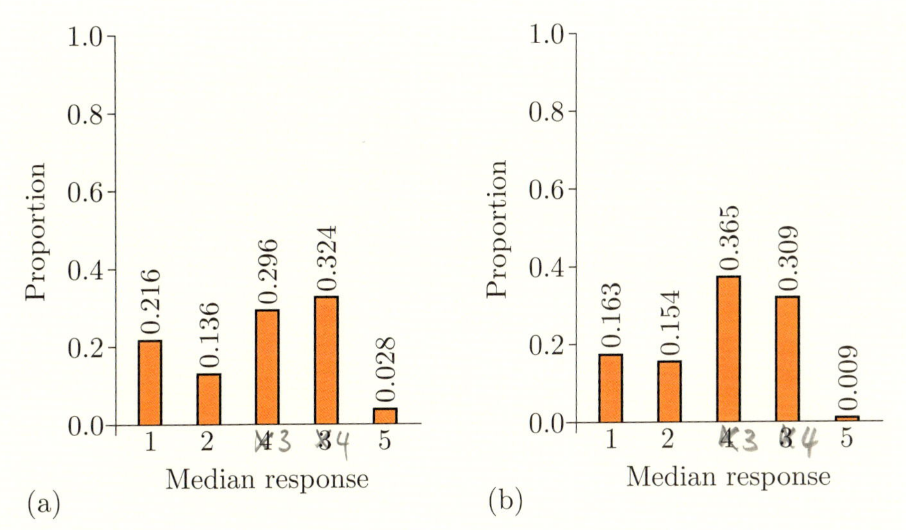

Figure 7 (a) Approximate proportion of samples of size three with each median response; (b) Approximate proportion of samples of size five with each median response

The pattern in Figure 7(a) is not very clear-cut. Not many of the samples have median 5; but one cannot say much more than that. In Activity 6 you found that the median response for the population as a whole was 3. Nearly one-third of the samples also had median 3 – but even more of them had median 4, and large numbers had median 1 or 2 as well. In Section 2 we found that larger samples tended to be more representative of the population. Is this true in terms of medians?

To investigate this, it is useful to have a similar description and picture of the median responses of all the samples of size five (and larger sample sizes). The picture corresponding to Figure 7(a) for the eight trillion (8 000 000 000 000) or so median responses of each of the samples of size five is shown in Figure 7(b).

The proportions here describe the distribution of the sample median for samples of size five. It tells us that about 0.163 of the samples of size five (i.e. 16.3%, or rather more than 1.3 trillion samples) have median response 1, about 0.154 of them have median response 2, about 0.365 of them have median response 3, about 0.309 of them have median response 4 and only about 0.009 of them have median response 5. This is another sampling distribution and it enables us to summarise very concisely all eight trillion samples of size five. Furthermore, it is precisely the type of summary picture we need to compare different sample sizes.

Comparing Figures 7(a) and 7(b), you can see, for instance, that a greater proportion of the samples of size five have a median of 3 (the population median response) than was the case for the samples of size three. How does the picture change as the sample size increases further?

You have now covered the material related to Screencast 3 for Unit 4 (see the M140 website).

3.4 Different sample sizes

Figure 8 contains pictures (corresponding to Figures 7(a) and 7(b) in Subsection 3.3) of the distributions of the sample median for several different sample sizes. For each sample size n there are a huge number of possible samples, each of which has a median, and the picture for sample size n shows the proportion of those medians which are 1, the proportion which are 2, the proportion which are 3, and so on.

Here, we use median as shorthand for median response.

Activity 8 *Effect of sample size*

Describe the most obvious change in the distributions in Figure 8 as the sample size n gets larger.

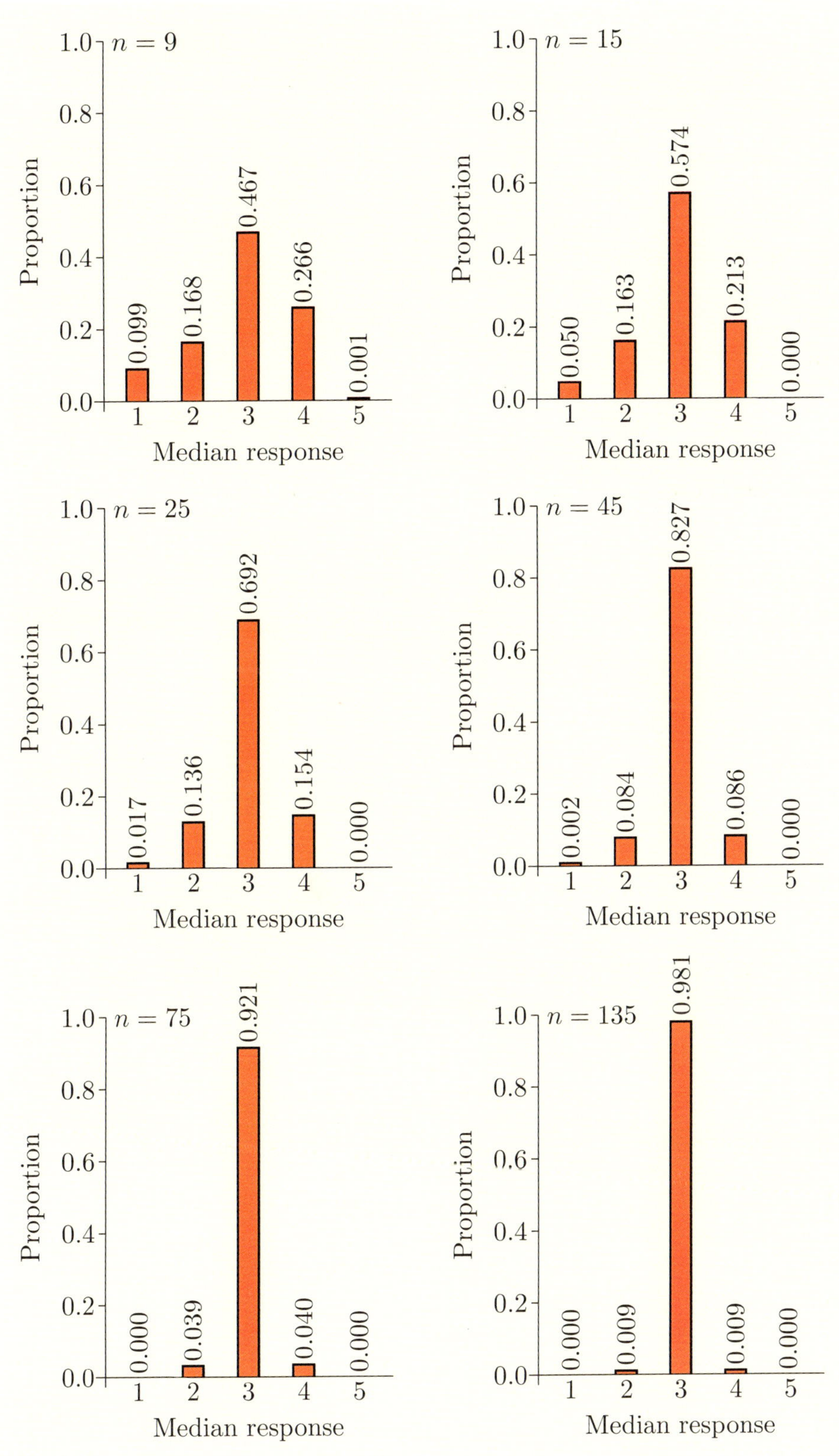

Figure 8 Approximate proportion of samples with each median response for various sample sizes

We have found that as the sample size increases the sample median becomes much more predictable and is much more likely to equal 3, which is the value of the population median. One important consequence of this is relevant to any investigation using samples, including those we considered in Sections 1 and 2.

If you choose a simple random sample of size five from the population described in Table 7, then you are, for example, more likely to choose one with median 3 than you are to choose one with median 5. This is because if you use simple random sampling, then each sample is equally likely to be chosen. You are therefore much more likely to choose one of the large number of samples with median 3 than one of the relatively much smaller number of samples with median 5.

If you choose a larger simple random sample, of size 15 say, then you are more likely to choose one with median 3 than you are to choose one with median not equal to 3; and if you choose a simple random sample of size 135, you are almost certain to choose one with median 3. Now there are an enormous number of possible samples of size 135 and *before* you choose one at random you have no idea which one will be chosen. However, you *can* nevertheless predict with reasonable confidence that its median will be 3. The larger the size of your random sample, the more certainly you can predict what its median will be.

The number of samples of size 135 is about 3×10^{170}: written out, this would be 3 followed by 170 zeros.

The patterns in Figures 7 and 8 can be described in words as follows. For all but the smallest sample sizes, the sample medians show a very clear and precise pattern: they are nearly all 3. As you found in Activity 6, the population median response is 3. Therefore, as the sample size gets larger, it becomes more and more likely that the sample median response will be the same as the population median response. In this precise sense, the pictures show that larger samples are more representative.

This type of pattern is very common. In general, patterns in sampling distributions from samples of different sizes show that larger samples are more representative. There is also usually a connection between patterns in the population values and patterns in collections of samples from that population (i.e. patterns in sampling distributions).

If, as here, we know the population values, then we can picture their distribution and thus see the patterns in them. The distribution could be pictured on a stemplot for small populations, but for a population of size 1000 this is not a very convenient picture. A common alternative is to use pictures like those used for the sampling distributions in Figures 7 and 8. As with the sampling distributions, we express each number in Table 7 as a proportion of 1000, the population size, and list these proportions on the picture. Thus Figure 9 is a picture of a **population distribution**. We shall study further pictures of population distributions in later units.

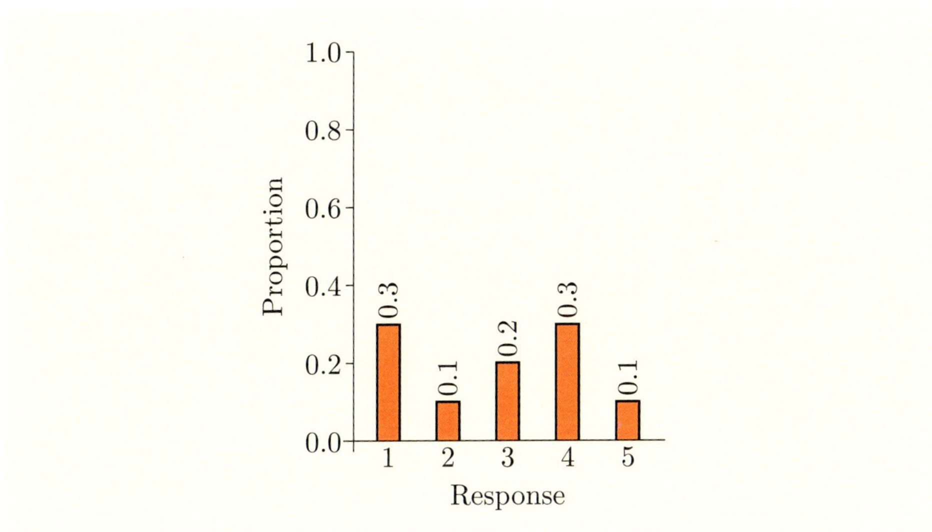

Figure 9 Proportion of members of population with each response

In statistics, interest often focuses on patterns that arise in the collection of all samples of a fixed size. These patterns lie behind many of the methods of analysing sample data that you will meet in later units. In the example we have been discussing, the patterns allowed us to say how likely it is that the sample median response is equal to the population median response. They could also tell us how close the sample median response is likely to be to the population median response; for example, for a sample size of 25 or above, the sample median response might be 2 or 4 (one away from the median) but is very unlikely to be 1 or 5 (two away). More generally, such patterns allow us to say how likely it is that a random sample will be representative in a particular sense, and they allow us to quantify how unrepresentative it is likely to be.

It is important to appreciate that these patterns *can* be described. This is done using sampling distributions. Pictures like those in Figures 7 and 8 are used to summarise sampling distributions and hence show patterns. They are also very useful for describing population distributions (as in Figure 9). So here are some activities based on the pictures in Figures 7 and 8.

Activity 9 *Most likely sample median*

For samples of size three (Figure 7), which value has the largest proportion of the median responses (i.e. what is the most likely median of a simple random sample of size three)?

Activity 10 *Sample median equals population median?*

For which of the sample sizes covered by these pictures (Figures 7 and 8) is it true that over 60% of the samples have median 3?

Another use for patterns of this kind is in choosing the sample size for a survey. Suppose that, for some reason, you were particularly interested in finding out the median of this population, on the basis of sample data. You could do this by finding the sample median and using it as an *estimate* of the population median. The patterns in Figures 7 and 8 show that this estimate would be fairly likely to be wrong if the sample size was only 3 or 5, but almost certain to be right if the sample size was 75 or 135. Such considerations would allow you to choose an appropriate sample size.

In this module there is not time to explain any further how to decide the size of sample which is needed for a particular survey, but one important point is that this does not depend greatly on the size of the target population. Figure 8 demonstrates that a sample of size 75 is very likely to lead to an accurate estimate of the median of a population of 1000 individuals whose responses follow the pattern shown in Figure 9. If the general pattern of responses for the population of the whole of the UK were similar to that shown in Figure 9, then a sample of size 75 would also be very likely to lead to an accurate estimate of the median response for the UK population, even though the UK population consists of well over 60 million individuals rather than 1000.

The most important general points that have been covered in this section are that the collection of all possible samples of a given size has a pattern, that some aspects of this pattern are very precise for all but the smallest sample sizes, and that in looking for such patterns it can be very useful to describe and picture distributions by expressing them in terms of proportions. The last two sections of this unit return to some practical matters involved in planning and running surveys.

You have now covered the material needed for Subsection 4.1 of the Computer Book.

Exercises on Section 3

Exercise 4 *Proportions for a sample of 9*

For sample size 9 (Figure 8, Subsection 3.4),

(a) approximately what proportion of the samples have median 1?

(b) approximately what proportion have median 2?

(c) approximately what proportion have median less than 3?

(d) approximately what proportion have median greater than 3?

Exercise 5 *A different population*

Suppose a different population of 1000 people gave the following responses:

Response	Rating	Number
Much worse off	1	200
Somewhat worse off	2	400
About the same	3	200
Somewhat better off	4	100
Much better off	5	100
Total		1000

(a) What is the median response for this population?

(b) Figures A, B and C show three distributions of a sample median. One is for a sample of size seven from the above population, one is for a sample of size 21 from the above population, and one is for a sample of size 21 from a different population. Giving your reasons, say which figure relates to which sample.

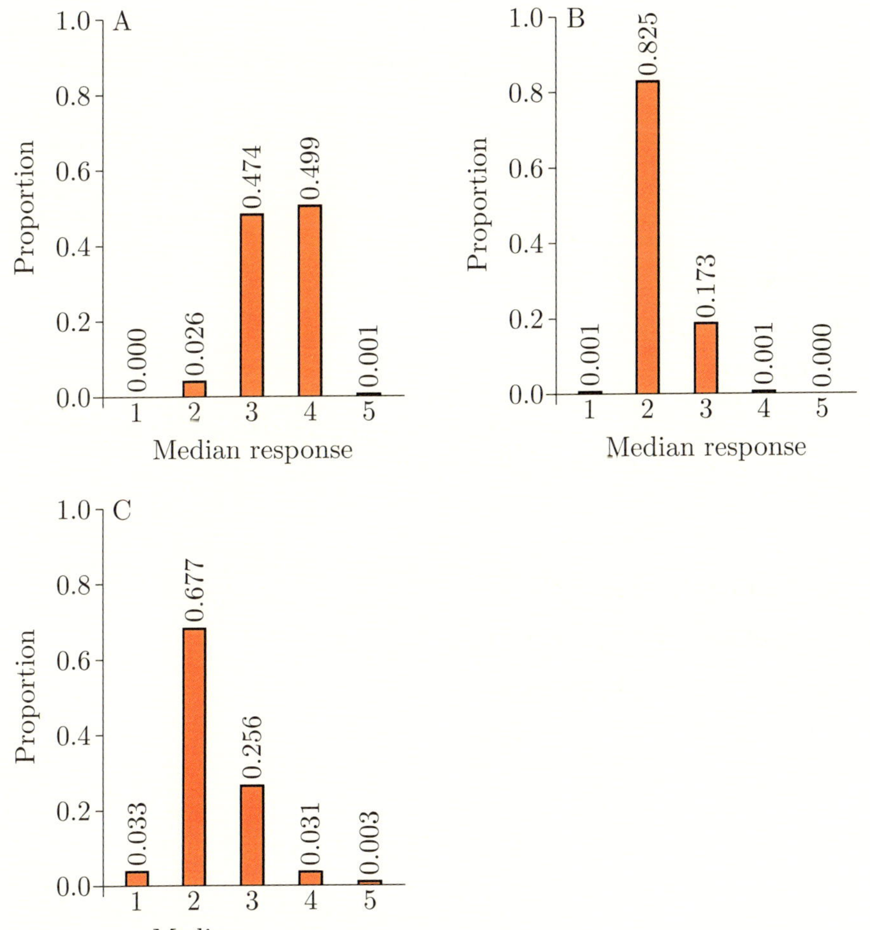

Figure 10 Distributions of three sample medians

4 More sampling methods

Section 2 of this unit introduced two ways of selecting a random sample for a survey – simple random sampling and systematic random sampling. In Section 4, more ways of choosing a sample for a survey will be introduced: stratified sampling, cluster sampling and quota sampling. Before this, in Subsection 4.1, you will learn about types of error that are associated with results obtained using survey data.

4.1 Types of error

If you intend to survey a population by investigating a random sample and inferring from data about this sample back to the population, then it is most unlikely that the results you get from the sample will be identical to those you would have got if you had obtained results from every individual in the population. For example, if you were interested in the mean, the mean of the sample will almost certainly not be the same as the mean of the population, although you hope that the two will not be very different. Statisticians refer to this difference as an **error** and there are several different types of error.

First, there is what is known as **sampling error**. As we saw in Section 3, different samples contain different individuals, and although there is a pattern in the possible results, we cannot know where our particular sample lies in the pattern. So there is variability due to sampling. This is the source of sampling error.

Second, there may be error introduced by using a poor sampling scheme. An example of this is a mobile phone survey where the sample is selected from a listing of mobile phone numbers. Selected people are contacted by phone. This survey has a **bias** in that people who do not own a mobile phone or who have chosen not to have their number listed could not possibly be included in a sample. A survey based on the electoral register would also include a bias against people who move house frequently. Another situation in which bias arises is *quota sampling*, which will be described in Subsection 4.5.

Third, there are other **non-sampling errors** which can arise from a variety of causes; for example, errors in recording responses or in transferring them to a computer, failure to contact individuals who are supposed to be included in a sample or refusal of people to cooperate with the interviewer.

Both the second and third types of error can be reduced or eliminated by planning the survey properly, by employing experienced interviewers and by careful checking. It is impossible to eliminate the first type, the sampling error, because this is inherent in the process of sampling. However, design of the survey can reduce the sampling error, as we shall see in this section.

Other things being equal, a larger sample size gives more accurate results but also leads to higher costs. In an ideal world with no resource constraints, sampling error could be eliminated completely by investigating the whole target population. However, in the real world the costs of collecting reliable data are considerable, so survey planning must involve careful consideration of the resources available.

The Gallup Poll and George Horace Gallup

George Gallup (1901–1984) made important advances in survey sampling methods and founded his own polling company (which became the *Gallup Organization*) in 1935. The company came to prominence the following year when it used a survey of 50 000 respondents to correctly forecast that Franklin Roosevelt would defeat Alf Landon for the U.S. presidency. An influential magazine at the time, the *Literary Digest*, conducted a much larger survey but incorrectly predicted that Landon would win. Moreover, Gallup's company correctly forecast the prediction that the *Literary Digest* would make, by following the sampling procedure they used, though with a much smaller sample size. The *Literary Digest* had sampled a list of its own subscribers and lists of car owners and telephone users, so that (in 1936) it was only sampling from the more affluent sections of the U.S. population, making its sample unrepresentative.

The *Gallup Poll* (one division of the Gallup Organization) conducts opinion polls in over 140 countries on an enormous range of political, economic and social issues. Its low point was probably in 1948, when it incorrectly forecast that Thomas Dewey would beat Harry S. Truman by a big margin in the U.S. presidential election. George Gallup believed the inaccuracy stemmed from ending his survey more than three weeks before the election.

George Gallup (1901–1984)

The aims of survey planning are to minimise both *costs* and *errors* (both sampling errors and non-sampling errors). These requirements are in conflict. Sampling error is reduced by choosing a larger sample, but costs are increased. We shall now briefly describe two further important tools of the survey planner's trade: first, a method of reducing sampling error (*stratified sampling*) and, second, a method of reducing costs (*cluster sampling*).

Despite producing results with no margin of error, the Eden poll is now defunct.

4.2 Stratified sampling

A method of sampling that reduces sampling error is often called **efficient**. This does *not* mean that it is cheap – such methods often cost more.

To reduce sampling error we have to reduce the potential variation between the different possible samples that we can choose. In other words, we want to make it more likely that the sample we choose is representative.

In Section 2 we assessed the representativeness of samples chosen from a listing of staff in the Sampling Department by analysing them with respect to gender and occupation. This was done by dividing the members of the sample into eight categories: these categories were the four occupational groups, each split into two genders. Having divided the sample into these eight categories, we then saw how the proportion of the sample in each category compared with the corresponding proportion for the whole population.

There were two reasons for using these particular eight categories for this analysis.

1. We knew the proportion of the whole population in each of these categories. (We could not base categories on salary levels, for instance, because they were not recorded on the list of the population that we had.)
2. It appeared likely that these categories were related to the subject of the investigation. To be more precise, it appeared that the data we collected from an individual on their economic well-being would depend on that individual's occupation and gender. It would not be possible to tell for certain if a sample was representative in terms of economic well-being without knowing the economic well-being of all the individuals in the population; and if we knew that, there would be no need to carry out the sample survey. But because economic well-being is thought to be related to occupation and gender, a sample that is representative in terms of occupation and gender is likely to be representative in terms of economic well-being too.

Categorising the population in this way is known as **stratification**: the eight categories are the **strata**. (A single category is a **stratum**.)

It is quite straightforward to ensure that any sample you might choose from the department is representative with respect to these eight strata. Instead of selecting members of the sample at random from the whole population, you would list the members of each stratum separately and then from each stratum select a number of individuals by simple or systematic random sampling. The selected individuals from a stratum form a **subsample**. You would then combine these subsamples (one subsample from each stratum) to get a sample from the whole population. This sample is then bound to be representative with respect to the strata, and is thus likely to be representative with respect to the subject of the investigation. Ideally, all the individuals in each stratum would be very similar to each other, so that whoever was selected from a stratum would be representative of that stratum. Then there would be comparatively little sampling error. A sample chosen in this way is a **stratified sample**.

This description of stratified sampling ignores one important point: how many individuals should be selected from each stratum, i.e. what should be the sizes of the subsamples? For example, suppose you want to deduce information about the average income of the members of the department (listed in Table 1) from data about the incomes of a sample. With a very small sample, there would not be much possibility of choice. With a sample of total size eight, you would have to choose a subsample of size one from each of the strata as it is an essential criterion of stratified sampling that in the sample there should be at least one member from each stratum.

However, if you are prepared to select a slightly larger sample, the ideas from Section 2 suggest that you should select approximately the same proportion of individuals from each stratum. If you wanted a sample of size 20 from the 86 members of the department, then you would select about the same proportion, 20/86, of the people in each stratum.

For example, there are 17 people in the stratum of female secretarial staff; you might select about $20/86 \times 17$, which is about four, people from this category, and you might select ten or eleven men from the male professional category. You would still have to select the single male administrator and the male secretary.

Stratum subsample size

If approximately the same proportion of individuals are to be selected from each stratum, then

$$\text{stratum subsample size} \simeq \frac{\text{sample size} \times \text{stratum size}}{\text{total population size}}.$$

(Any subsample size less than one would be set equal to one.)

As described in Section 2, if you began by listing the population in order of strata (all the female professionals, then all the male professionals, followed by all the female administrators, the male administrator, and so on) and then chose a systematic random sample from the whole list, then the subsample sizes within each stratum would automatically come out to be approximately proportional to the stratum sizes.

However, when a little more is known about the population, it is sometimes better not to select a stratified sample in which the subsample sizes are proportional to the stratum sizes. For example, if you had the extra information that the incomes of the male professionals have a much larger spread than those of the female secretarial staff, then it would be better *not* to select the same proportion of each of these strata. This is because you need to obtain more information about the stratum with the larger spread in order to get the same amount of accuracy in your results. You should therefore choose a larger subsample from such a stratum, i.e. you should choose a relatively larger proportion of *male professionals* and a relatively smaller proportion of *female secretarial staff*.

This procedure makes more sense when we are thinking about sampling a large population, like electors in the UK, rather than a department with 86 people. With a large population, there would be thousands of people in each stratum, and it is easy to consider drawing subsamples whose size is proportional to the stratum size, or perhaps varying the proportions to take account of other available information.

In practice, most surveys that use subsample sizes which are not proportional to stratum sizes have a different reason for doing so. Suppose you were planning a survey of the adult population of England and Wales to investigate their subjective feelings on how well off they are. You would probably want to use stratified sampling, and you might well choose to stratify according to region of residence. You might work out that a total sample size of, say, 2000 would allow you to estimate sufficiently accurately what you want to know about the population of England and Wales as a whole. However, you might be particularly interested in comparing the results for Greater London with those for the rest of the country. Roughly one-seventh of the population of England and Wales lives in Greater London, so if subsample sizes were chosen in proportion to stratum sizes, the Greater London subsample would consist of under 300 individuals. Such a sample size would probably not allow you to estimate sufficiently accurately what you want to know about the population of Greater London. You might therefore decide to increase the sample size for the Greater London subsample. In general, subsample sizes are often chosen so that appropriately accurate information is available on strata of particular interest, as well as for the population as a whole.

Stratification

Stratification is the categorisation of the population into strata that are:

- exhaustive: every member of the population must belong to a stratum
- mutually exclusive: no member of the population can belong to more than one stratum
- relevant to the subject under investigation: within each stratum, individuals should as far as possible be similar with respect to this subject
- known for all population members before the sample is chosen: otherwise a list of the individuals in a stratum from which to choose the subsample would not be available.

A stratified sample might then be chosen by selecting approximately the same proportion of individuals from each stratum. Such a stratified sample will be representative of the population with respect to the sizes of these strata. However, a stratified sample need not be chosen in this way, and often further knowledge about the population or the purpose of sampling will suggest better methods of selecting individuals from the strata.

These methods of stratified sampling ensure that the patterns in a stratified sample are less likely to be different from those in the population than are the patterns in a simple random sample of the same size. Therefore, the use of a stratified sample leads to more reliable results than the use of a simple random sample of the same size; in other words, the sampling error is reduced.

Example 4 *Survey of consumer prices*

You may remember from Unit 2 (Section 5) that the calculation of the RPI uses a monthly survey of retail prices carried out by a market research company on behalf of the UK Office for National Statistics. In this survey, prices are collected from a sample of shops situated in approximately 150 locations across the UK. This sample of shops is stratified: each shop is put into one stratum according to which of the 12 regions of the country it is in and which of the three types of retail outlet it is.

Activity 11 *Who bought the seed?*

Suppose that you work for a mail order seed company and you wish to carry out a sample survey of the population of UK customers who bought seed of a new variety of pea to find out their opinion of it. You have computerised records of the names and addresses of all these customers, and of the amount of seed of this variety that each of them bought. How would you go about dividing this population into strata?

Stratified sampling has one disadvantage which is normally relatively minor: it can increase costs. This is because to use this method it is necessary to spend time discovering information about the population and then carefully distinguishing the strata and deciding the subsample sizes. We shall now look at a method which, in contrast, can produce dramatic savings in costs in certain types of survey.

4.3 Cluster sampling

Many surveys involve interviewers contacting individual members of the chosen sample in their homes or work places. A survey of this kind can be enormously expensive, particularly if it covers a wide geographical area such as the whole of the UK, because the interviewers' travel time and transport costs are both considerable. It is obviously in the interests of economy to arrange, if possible, for the individual members of the sample to be not too widely dispersed geographically.

One major survey that involves such personal interviews is the Living Costs and Food Survey.

Here, then, is a brief description of **cluster sampling**: a method that cuts the costs of such surveys by restricting the sample to a limited number of geographical areas.

Choosing a cluster sample

Cluster sampling works as follows:

1. Find suitable geographical areas.
2. Choose, preferably using random methods, a limited number of these geographical areas.
3. For each of these chosen geographical areas, choose a subsample from those members of the population in that area.
4. Combine these subsamples (one from each chosen area) to get a sample.

The population in each geographical area is a **cluster**, and such a sample is a **cluster sample**. Clusters may also consist of entities other than geographical areas.

For this method of cluster sampling to produce representative samples, it is essential that the populations in the chosen clusters are, between them, representative of the whole target population.

In 1947, Hollywood made a film (*Magic Town*, starring James Stewart) about a small town in the Midwest of the USA which was a microcosm of American Society. This single town of about 2000 inhabitants was found to represent the whole country in its social, economic and political characteristics. Any such town, in any country, would be ideal for official surveys, for market research and for public opinion polls because all such surveys could confine their attention to a sample from this one town, i.e. they could choose just one cluster. A few hours' work interviewing a random sample of individuals from this town would produce representative results about the whole population of the country, saving enormous amounts of time and money. Such towns, however, exist only in a Hollywood producer's imagination. The real world is no Hollywood! Towns within a country differ quite a lot in their characteristics, depending, for example, upon where they are, the age of their populations and the major local employers.

For this reason, it is never sensible to confine a cluster sample to a single cluster. The usual practice is to choose several clusters using random sampling; then a subsample is selected from each chosen cluster, again normally by simple random sampling.

There are various forms of cluster sampling. One form is described below.

One form of cluster sampling

1. Specify the number of clusters to use in the survey and the proportion that is to be surveyed from each of the selected clusters.
2. Choose which clusters to use at random, with each cluster having the same probability of being included in the survey.
3. Draw a simple random sample from each of these clusters. The clusters may differ in their sizes, and the sizes of the subsamples drawn from them should vary correspondingly: subsample approximately the same pre-specified proportion of each cluster.

A desirable property held by this form of cluster sampling is that every individual in the target population has approximately the same probability of being included in the survey. (Small differences between the probabilities will usually be inevitable because the sample sizes must be whole numbers.) A drawback, though, is that the total sample size will partly depend on which clusters are chosen – if large clusters are chosen by chance in step (b), then the total sample size will be larger than when step (b) yields small clusters. There are forms of cluster sampling that avoid this drawback, but we will not consider them in M140.

Although cluster sampling saves money, it also has a disadvantage: other things being equal, cluster sampling will almost always lead to greater sampling errors than would arise in a simple random sample of the same size. The reason for this is that individuals within clusters tend to be less variable than individuals in the target population as a whole. Two people living in the same town are likely to be more similar than two people living in different towns. By restricting the sample to the chosen clusters, it is thus likely to be less representative.

There are circumstances in which cluster sampling is likely to produce a sample that is more representative than a simple random sample of the same size, but in practice these circumstances hardly ever arise.

However, suppose a survey is being planned within a fixed budget. Very often the cost savings achieved by using clustering allow the sample size to be increased to such an extent that the results from the cluster sample are considerably more reliable than the results would be from the very much smaller unclustered sample that could be afforded.

Do not forget that this argument applies only to surveys using interviewers who have to travel. It would not apply, for example, in a survey carried out by post. For many such postal surveys, there is no reason for using clustering on a geographical basis. However, there is another good reason for using cluster sampling in some situations. To draw a simple random sample, a complete list of the target population is required. For some populations, it would be a major undertaking to produce such a list. No complete single listing of all UK schoolchildren exists, for instance, and it would not be feasible to produce one. It would be much more feasible, for a survey of this population, to obtain a list of all schools, to choose a limited number of schools as clusters, make a list of the pupils in each of the selected schools, and draw samples from these lists.

Although cluster sampling usually makes use of geographical areas, there are other ways of dividing a population into clusters. For example, suppose a chocolate manufacturer wanted to sample his chocolates at the end of production, in order to test for quality. It would be economical to select boxes of chocolates at random and then to select several (or perhaps, all) of the chocolates from the selected boxes for testing. This would avoid wasting too many boxes. Here, each box of chocolates is a cluster.

4.4 Stratified and cluster sampling

Let us now summarise the main points from the last three subsections and compare these two methods.

Stratified sampling	**Cluster sampling**
• Each stratum focuses on one section of the population, such as those of a specified gender in a particular age group.	• Each cluster should be, as far as possible, a representative cross-section of the whole population.
• Every member of the population must be in one and only one stratum.	• Every member of the population must be in one and only one cluster.
• A stratified sample includes members of every stratum.	• A cluster sample excludes all the members of some (usually most) of the clusters.
• Stratified sampling decreases sampling error compared to a simple random sample of the same size (i.e. it is more efficient) but slightly increases costs.	• Cluster sampling often decreases costs but usually increases sampling error compared to a simple random sample of the same size (i.e. it is less efficient).

Many well-planned surveys use both strata and clusters. An example of such a survey is the Living Costs and Food Survey, introduced in Unit 2. There are also elements of both in quota sampling, as you will see in the next subsection.

4.5 Quota sampling

Quota sampling is a procedure that is used frequently for market research surveys and opinion polls. Firstly the sample size is determined (usually by consideration of costs), and then each interviewer is allocated a quota of interviews to achieve. The interviewers are then sent out to contact suitable respondents at selected sites in selected towns (Figure 11).

Figure 11 Data collection

These sites might be supermarkets, railway stations, high streets, etc. Thus the quota sample is a cluster sample. The sample is stratified by requiring interviewers to interview a fixed number of people in specific groups such as age, gender and occupation groups.

A quota sample is not a random sample: the selection of individuals is haphazard rather than random.

Quota sampling is economical because it produces quick results. These results are, however, often of dubious reliability because the method can introduce error. Market researchers are fond of quoting the apocryphal story of the interviewer who quickly achieved his full quota of interviews from people queuing for a train at Liverpool Street Station in London. The survey was about gambling and all those interviewed were waiting for a special train to take them to the Newmarket horse races!

Another disadvantage of quota sampling is that it is usually difficult to give a numerical estimate for how unrepresentative the results are likely to be. It is possible to give such estimates for random sampling methods, using the ideas of probability that you will meet in Unit 6.

4.6 Sampling from the electoral register

Most of the methods of choosing a sample described in this unit require a list of all the individuals in the target population. This list is sometimes called the **sampling frame**. One sampling frame that has commonly been used in the UK for surveys of individual adults and of households is the **electoral register** (such as that shown in Figure 12). This lists all electors and it is possible to buy an edited version. The full register contains almost all adults who are eligible to vote, as the registration of eligible voters is compulsory in the UK. However it does not contain many non-EU citizens or any people aged under 17. (People can be registered to vote from age 17, though their registration is not activated until they reach their 18th birthday.) Also, the edited register does not include anybody who has chosen not to be included in the edited version. Another drawback of the electoral register is that it is out-of-date even when it is first published, because compiling a relatively complete list of a large human population is time-consuming.

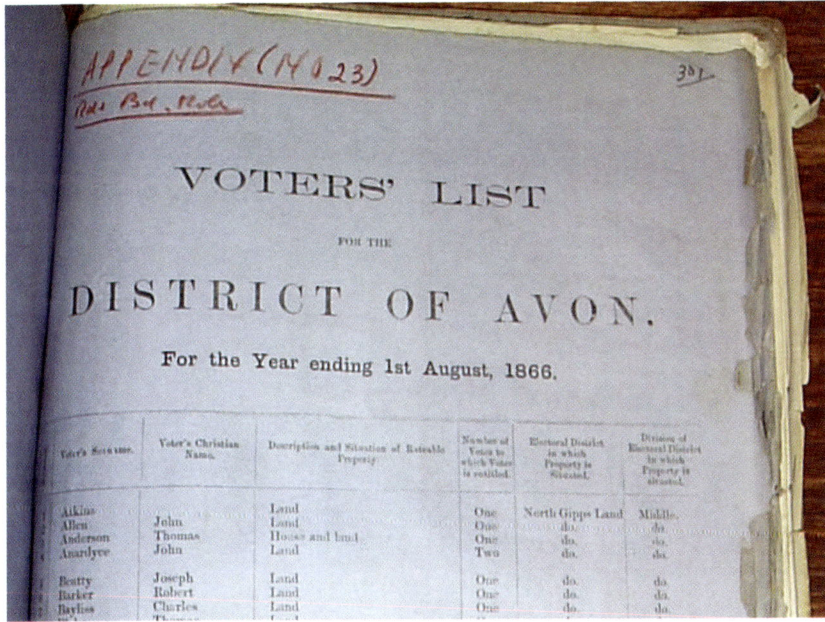

VOTERS' LIST

FOR THE

DISTRICT OF AVON.

For the Year ending 1st August, 1866.

Voter's Surname.	Voter's Christian Name.	Description and Situation of Rateable Property.	Number of Votes to which Voter is entitled.	Electoral District in which Property is Situated.	Division of Electoral District in which Property is situated.
Aikins		Land	One	North Gipps Land	Middle.
Allen	John	Land	One	do.	do.
Anderson	Thomas	House and land	One	do.	do.
Anardyce	John	Land	Two	do.	do.
Beatty	Joseph	Land	One	do.	do.
Barker	Robert	Land	One	do.	do.
Bayliss	Charles	Land	One	do.	do.

Figure 12 Avon Roll 1866

We will use the electoral register for a part of Milton Keynes to illustrate some of the survey methods that have been discussed. We will suppose the target population is the adult residents of five streets (Jersey Close, Kerrera Close, Lytham Gardens, Melton Gardens and Norfolk Place) and that the purpose of the survey is to learn about their bus usage. People participating in the survey will be asked:

Did you use a bus service in Milton Keynes in the last week?

Table 11 lists the adults living in the target streets, based on the electoral register. It also records whether or not they had used a bus service in Milton Keynes during the week – though this information would only be known for those people questioned in the survey.

Table 11 Bus usage in a part of Milton Keynes

Registration number	Name	Street number	Bus user?
Jersey Close			
977	Denton, George	1	Y
978	Wells, Joan F	2	Y
979	Hanrahan, Brian K	3	N
980	No Elector	4	
981	Jones, Ian	5	Y
982	Jones, Linda	5	Y
983	Abbott, David	6	N
984	Abbott, Mary R	6	Y
985	Donegan, Andrew B	7	N
986	Donegan, Margaret H	7	N
987	Turner, Thomas J F	8	Y
988	Turner, Florence P	8	Y
989	West, Michael J	9	Y
990	West, Jean P	9	N
991	Nelson, Sheila A	9	N
992	Mason, Arthur B	10	N
993	Mason, Joan M	10	Y
994	Wilson, Annabel N	11	N
995	Wilson, Lillian	11	Y
996	Chapman, Reginald R	12	Y
997	Chapman, Iris	12	Y
998	Watson, Richard T	12	N
999	No Elector	13	
1000	Mercer, Gladys C	14	Y
Kerrera Close			
1001	Groves, Jacqueline F	1	Y
1002	Drinkwater, James G	1	Y
1003	Tong, Michael	2	N
1004	Burton, Christopher N	3	Y
1005	Hexton, Amara	4	N
1006	Hexton, John	4	Y
1007	Smith, Alan C	5	Y
1008	Dixon, Mary C	6	Y
1009	Daly, Sean	6	Y
1010	Ho, Audrey	7	N
1011	Tongwell, Kim	8	N
1012	Clark, Michael E	9	N
1013	Clark, Jennifer	9	Y
1014	Christon, John E	10	Y
1015	Christon, Clare M	10	Y
1016	Dunn, Garry A	11	Y
1017	Dunn, Mary E	11	Y
1018	Edwards, Kathleen	12	N
1019	Edwards, Vince L	12	Y
1020	Price, Eleanor T	12	N
1021	Goulding, Matthew M	13	Y
1022	Goulding, Janet	13	Y
1023	Turner, Lee	14	Y
1024	Bailey, Ivy W	15	Y
1025	McCann, Raymond D	16	Y
1026	McCann, Victoria K	16	Y
1027	Wyatt, Edith	17	N

Registration number	Name	Street number	Bus user?
Lytham Gardens			
1028	Kerr, John M B	1	*N*
1029	Kerr, Susan	1	*N*
1030	Kerr, Lynn	1	*Y*
1031	Kerr, David	1	*Y*
1032	Kohler, Martina	2	*Y*
1033	Kohler, Nicholas	2	*N*
1034	Clements, Neil S	3	*N*
1035	Clements, Marie A	3	*N*
1036	Clements, Ian P	3	*N*
1037	Patel, Suresh	4	*Y*
1038	Knight, Patricia H	5	*N*
1039	Bolton, Samuel T	6	*N*
Melton Gardens			
1040	Clarke, David P	1	*N*
1041	Clarke, Annette M L	1	*N*
1042	Barnard, Ruby	2	*N*
1043	No Elector	3	
1044	French, Richard E	4	*N*
1045	Coe, Alanah	4	*Y*
1046	Smith, Angela	5	*Y*
1047	Ferguson, Brian	6	*N*
1048	Ferguson, Sally	6	*N*
1049	Ferguson Michael	6	*Y*
1050	Shah, Jaya	7	*Y*
1051	O'Neill, Thomas	8	*Y*
1052	O'Neill, Mary S	8	*Y*
1053	Hedley, Robert M	8	*N*
1054	Scott, Ian R	9	*Y*
1055	Scott, Dorothy G	9	*N*
1056	McGregor, David E	10	*N*
1057	McGregor, Aileen J	10	*N*
1058	Paine, Darrell R	11	*N*
1059	Paine, Lynne C	11	*N*
Norfolk Place			
1060	Fisk, Catherine A	1	*N*
1061	Hatley, Brian J	2	*N*
1062	Brooke, Denise	2	*Y*
1063	Lang, Deborah M	3	*Y*
1064	Flynn, Horace I	4	*N*
1065	Flynn, Ann C	4	*Y*
1066	Shah, Dipak	5	*Y*
1067	Shah, Mala	5	*N*
1068	McTaggart, William E	6	*Y*
1069	McTaggart, Christine V	6	*N*
1070	McTaggart, James J	6	*N*
1071	Hall, Stephen D	7	*N*
1072	Godman, Janet K	7	*N*
1073	Weston, Zoe	8	*N*
1074	Uttley, Muriel O	9	*Y*

To sample from the electoral list in Table 11, we use random numbers and relate these to the registration numbers. The registration numbers run from 977 to 1074, so, with random number tables, it is efficient to use pairs of random digits:

- 79 would mean 'Registration number 979'
- 73 would mean 'Registration number 1073'
- 00 would mean 'Registration number 1000'.

We would ignore 75 and 76, and also the pairs corresponding to 'No Elector'.

Example 5 *Simple random sample of 12 electors*

Suppose a simple random sample of twelve electors is required. If we use the random number table in the appendix and start at the beginning of row **49**, then the selected random numbers are:

$$96, 00, 26, 82, 60, 22, 02, 60, 69, 99, 09, 67, 01, 12, 01, \ldots$$

Equating these to the corresponding electoral registration numbers determines our sample. (The second 60 will be ignored, because we want no repeats, and 99 will be ignored because 999 is a 'No Elector'.) The electors in the sample and their characteristics are given in Table 12.

Table 12 Bus-usage in a simple random sample

Registration number	Name	Address	Bus user?
996	Chapman, Reginald R	12 Jersey Close	*Y*
1000	Mercer, Gladys C	14 Jersey Close	*Y*
1026	McCann, Victoria K	16 Kerrera Close	*Y*
982	Jones, Linda	5 Jersey Close	*Y*
1060	Fisk, Catherine A	1 Norfolk Place	*N*
1022	Goulding, Janet	13 Kerrera Close	*Y*
1002	Drinkwater, James G	1 Kerrera Close	*Y*
1069	McTaggart, Christine V	6 Norfolk Place	*N*
1009	Daly, Sean	6 Kerrera Close	*Y*
1067	Shah, Mala	5 Norfolk Place	*N*
1001	Groves, Jacqueline F	1 Kerrera Close	*Y*
1012	Clark, Michael E	9 Kerrera Close	*N*

Eight individuals in this sample of 12 people are bus users, so the sample estimate of the percentage of bus users in the population is

$$\frac{8}{12} \times 100\% \simeq 66.7\%.$$

In the target population of 95 electors, there are actually 49 people who used the bus in the previous week, so the true percentage of bus users is $49/95 \simeq 51.6\%$.

Activity 12 *Systematic random sample*

Suppose a systematic random sample of about one-eighth of the targeted electors is required. Select such a sample, taking the first digit in the range 1 to 8 from row **6** as a random start. List the names of the electors in the sample and whether they are bus users. Based on this sample, what is the estimated percentage of bus users in the target population?

Some sampling schemes divide the population into categories that are sampled separately. (Some categories might not be sampled, as in cluster sampling, for example, where only selected clusters are sampled.) Having chosen the categories to sample, each category is taken in turn and a simple random sample drawn from it.

Example 6 *Stratified sample of 12 electors from two strata*

Suppose the five streets that give our target population can be sensibly divided into two strata: Jersey Close and Kerrera Close were, at the time of the survey, both newly built and form one stratum, while Lytham Gardens, Melton Gardens and Norfolk Place were all built about twenty years earlier and form a second stratum. The strata are of similar size (49 electors in one stratum and 46 in the other), so we will sample the same number of people from each stratum, i.e. six from each.

We will start in row **16** of the random number table.

16 47 14 97 61 57 30 93 88 12 88 58 15 75 17 45
17 98 75 58 14 05 05 16 72 57 34 20 46 91 04 44
18 64 71 77 50 51 00 61 02 60 51 13 61 34 33 73.

The electoral registration numbers for the first stratum range from 977 to 1027, so we look through the random numbers picking out those between 77 and 99, and those between 00 and 27, but ignore duplicates and those corresponding to 'No Elector'. This gives 14, 97, 93, 88, 12 and 15. For the second stratum, we start reading random numbers from where the previous sample ended, picking out those between 28 and 74: 45, 58, 72, 57, 34 and 46. The electors corresponding to these numbers, together with their characteristics, are listed by stratum in Table 13.

Table 13 Bus usage in a stratified sample

Registration number	Name	Address	Bus user?
Jersey Close and Kerrera Close			
1014	Christon, John E	10 Kerrera Close	Y
997	Chapman, Iris	12 Jersey Close	Y
993	Mason, Joan M	10 Jersey Close	Y
988	Turner, Florence P	8 Jersey Close	Y
1012	Clark, Michael E	9 Kerrera Close	N
1015	Christon, Clare M	10 Kerrera Close	Y
Lytham Gardens, Melton Gardens and Norfolk Place			
1045	Coe, Alanah	4 Melton Gardens	Y
1058	Paine, Darrell R	11 Melton Gardens	N
1072	Godman, Janet K	7 Norfolk Place	N
1057	McGregor, Aileen J	10 Melton Gardens	N
1034	Clements, Neil S	3 Lytham Gardens	N
1046	Smith, Angela	5 Melton Gardens	Y

Seven individuals in this sample of 12 people are bus users, so this sample estimates the percentage of bus users in the population as

$$\frac{7}{12} \times 100\% \simeq 58.3\%.$$

Example 6 is the subject of Screencast 4 for Unit 4 (see the M140 website).

Activity 13 *Cluster sampling with subsamples of one-third*

Suppose that the streets in the population listed in Table 11 were widely separated geographically, and that therefore you wanted to use cluster sampling for your survey, restricting your sample to just two of the streets and sampling approximately one-third of the individuals in each cluster. Obtain the sample using the following procedure:

- Number the streets from 1 to 5 in the order in which they are listed. Using *single* random digits, and starting at the beginning of row **26** of the random number table in the appendix, select the two streets to be sampled. These streets are to be sampled in the order in which they are selected.
- Determine the sizes of the samples to take from each cluster (street) by dividing each cluster size by 3 and rounding the results *up* to whole numbers.
- To select individuals for the subsample from the first selected street, use pairs of digits starting at row **82** of the random number table. No person may be selected more than once. To select individuals from the second subsample, continue from the point reached in the random number table after selecting the first subsample, and apply the same procedure again.

List the people chosen in the subsamples and estimate the proportion of bus users in the target population.

Activity 13 is the subject of Screencast 5 for Unit 4 (see the M140 website).

4.7 Some more considerations

Even if you ever thought that sampling would be child's play, you should now be able to appreciate that it is a good deal more difficult than pulling rabbits out of hats, and in addition, that it can involve a lot of hard slog. Here are a few more of the problems that abound in this work.

- **Defining the target population.** Sometimes this is not straightforward. For example, in an opinion poll designed to predict the result of an election, the target population is all those people who will actually vote on polling day, but who these people are cannot be known beforehand.
- **Listing the target population.** Most of the methods of choosing a sample described in this unit require a sampling frame. (An advantage of cluster sampling is that it does not require a full sampling frame.) It is often difficult to obtain an accurate list, as you saw in the description of sampling from the electoral register.
- **Non-contact and non-response.** Often it is impossible to contact everyone in the sample, and some of the individuals contacted may not be able or willing to provide the required information.
- **Questionnaire design.** This could well be the subject of a whole unit. Devising questions that will discover the required information is not easy. Also, for example, the way in which the questions are asked by the interviewer may well affect the answer.
- **Clerical errors.** No matter how carefully the work is done there are certain to be errors in recording and transcribing the data. Many of these will, however, be discovered if the data are analysed sensibly.

In this section, you have read about the principles involved in cluster sampling, stratified sampling and quota sampling. You now know about some of the problems in sampling, and in particular some problems of sampling from the electoral register.

Exercises on Section 4

These exercises consider how sampling might be used to investigate households whose expenditure may not fit typical patterns used by the Retail Prices Index (RPI).

Exercise 6 *Cluster sampling?*

Households which own their home outright, and therefore do not make either mortgage or rent payments, might well have a considerably different expenditure pattern to other households, and the RPI may therefore not be an accurate indicator of inflation as they experience it, particularly as the Housing sub-group has the highest weight in the RPI. Suppose you are required to select a national sample of such households so that their expenditure can be analysed separately.

(a) State, with a reason, whether cluster sampling would be a valid and appropriate method to use for the initial stage of selecting such a sample.

(b) Explain which method of sampling you would use to select the individual households in your final sample, justifying your choice of method.

Exercise 7 *Sampling methods and sampling frames*

The Motoring Expenditure sub-group has the second-highest weight in the RPI. In some rural areas, households which do not own a motor vehicle, and are therefore dependent on public transport, may have a different expenditure pattern to the majority of households that do own a vehicle. The RPI may therefore not be an accurate indicator of inflation as experienced by rural households without a vehicle. Suppose you are required to select a national sample of such households so that their expenditure can be analysed separately.

(a) A pilot survey is to be carried out in one area. What official records might you want to access to obtain a suitable sampling frame from which a sample of such households could be obtained?

(b) State which sampling method you would use to select the sample from the sampling frame, justifying your choice.

Exercise 8 *Stratified sampling*

Suppose the electorate given in Table 11 divides into three strata: Jersey Close, Kerrera Close and the other three roads. A random sample of size 12 is to be drawn from this population using stratified random sampling.

(a) Select the subsample sizes so that they are approximately proportional to the stratum sizes, ensuring that the total sample size is 12.

(b) Select the sample, using simple random sampling from each stratum in turn. Start at the beginning of row **52** of the random number table in the appendix. Write down the names of the electors you select and whether or not they are bus users.

(c) Calculate the percentage of electors sampled who are bus users and comment briefly on how well your sample represents the target population (of all adults living in this part of Milton Keynes) in terms of using the bus service.

5 Computer work: sampling

In Section 3, you looked at sampling from a target population and learned about sampling distributions. In this section, you will explore the sampling distribution for samples of size 3 taken from a particular target population, followed by looking at sampling distributions for samples of different sizes. You will then learn how to use Minitab to produce simple random samples.

You should now turn to the Computer Book and work through Subsection 4.1, if you have not already done so, followed by the rest of Chapter 4.

Summary

This unit has focused on statistical issues surrounding one method of data collection – surveys. In a survey, information is collected about a sample of individuals and used to draw conclusions about the population as a whole. Different methods are used to select samples, the best method depending on the survey and the target population.

- In simple random sampling, every possible sample of a given size has an equal chance of being selected. This is usually done by selecting individuals at random from the population. In systematic random sampling, individuals are chosen by working systematically down a list, with only the starting point chosen at random.
- Stratified sampling and cluster sampling assume that the population can be split into groups. In stratified sampling, individuals from every group are selected, ensuring that every group is represented in the sample. In cluster sampling, individuals in the sample only come from selected groups, ensuring that sampling process is more cost-efficient.
- In quota sampling, individuals are not selected at random, though they are chosen so that different groups in the population are represented fairly.

You have also learned in this unit about the sampling distribution of the median. That is, how the sample median varies according to which particular sample happened to be selected. You have seen that the sample median is not necessarily equal to the population median, even when there are just five categories to choose from. Indeed when the sample size is very small, it might be more likely to be different to the population median. However as the sample size increases, it becomes more likely that the sample median is the same as the population median.

Learning outcomes

After working through this unit, you should be able to:

- explain in general terms why a well-chosen sample is an economic and accurate method of collecting data about a population
- choose a simple random sample using random numbers and a labelled list of the target population
- choose a systematic random sample using random numbers and a labelled list of the target population
- describe the differences between, and outline the relative strengths and weaknesses of, simple and systematic random sampling
- give an example of the type of pattern that can be seen in the collection of all possible samples of a given size
- interpret descriptions and pictures of distributions which are expressed in proportions
- describe the principles involved in cluster sampling and stratified sampling
- describe quota sampling in general terms
- choose a random sample for a stratified survey using random numbers and a labelled list of the target population
- choose a random sample for cluster sampling using random numbers and a labelled list of the target population
- describe some of the problems in sampling, and in particular some problems of sampling from the electoral register.

Appendix: random number table

This table contains 3000 random digits (i.e. throws of a ten-sided die labelled 0, 1, …, 9).

1	98 06 77	46 16 63	99 80 81	82 15 48	96 12 56
2	41 25 66	21 51 66	11 34 33	18 36 41	33 18 70
3	68 58 71	24 92 06	94 84 48	92 96 32	29 00 60
4	78 32 89	76 61 03	01 20 94	36 39 87	52 27 23
5	50 76 11	36 13 84	32 93 72	29 04 41	25 43 89
6	71 58 45	43 72 69	18 67 32	57 29 57	02 58 68
7	82 14 64	47 40 74	53 03 75	40 28 63	53 36 90
8	41 01 53	67 41 78	84 29 26	34 42 19	82 31 79
9	30 63 22	27 28 69	36 23 99	52 29 03	87 28 54
10	73 09 03	54 20 02	55 49 48	46 75 42	62 63 42
11	66 41 48	46 17 24	82 51 86	86 53 66	95 57 95
12	70 10 21	02 71 89	14 80 64	32 58 17	35 65 55
13	59 55 94	44 77 90	01 99 79	48 28 61	93 87 17
14	75 81 42	45 69 28	23 90 46	24 32 97	64 41 70
15	72 22 05	84 39 89	57 73 84	86 57 76	79 08 65
16	47 14 97	61 57 30	93 88 12	88 58 15	75 17 45
17	98 75 58	14 05 05	16 72 57	34 20 46	91 04 44
18	64 71 77	50 51 00	61 02 60	51 13 61	34 33 73
19	43 12 15	66 40 56	39 77 75	32 80 30	22 90 95
20	59 70 46	36 67 19	12 59 39	42 35 24	69 86 14
21	25 84 20	27 35 05	54 21 39	04 77 69	78 76 99
22	40 52 36	07 18 99	79 27 36	30 97 14	72 64 82
23	48 38 90	79 26 63	50 41 87	76 31 13	81 55 34
24	61 91 66	85 68 10	40 47 44	71 56 81	00 34 07
25	45 40 26	25 37 27	02 15 26	27 51 87	18 91 30
26	32 57 79	72 02 27	96 10 62	63 07 30	01 40 97
27	69 23 49	01 02 17	28 23 72	71 46 39	24 46 39
28	63 80 25	47 36 69	73 39 21	23 93 10	09 50 45
29	31 30 49	19 65 12	33 87 76	64 22 62	66 61 88
30	68 42 66	14 60 63	24 06 92	94 21 52	71 37 19
31	52 77 76	33 55 75	78 03 11	18 04 23	12 72 46
32	19 05 93	62 41 96	47 15 34	80 17 23	06 44 75
33	15 23 16	85 63 28	62 03 72	11 74 17	35 37 09
34	32 84 18	60 89 57	09 25 31	82 79 92	10 08 71
35	59 10 86	85 92 14	14 17 38	59 35 24	12 53 88
36	18 56 17	74 42 45	19 35 75	18 37 47	42 78 08
37	28 87 01	51 67 42	00 77 30	16 31 06	67 42 75
38	83 25 37	02 91 92	05 16 09	07 35 84	59 15 44
39	12 09 73	08 61 72	89 23 91	85 76 99	29 55 48
40	64 74 95	68 36 68	69 99 56	33 78 08	84 31 87
41	77 46 18	83 52 40	05 76 20	95 40 64	73 67 44
42	06 69 75	42 75 68	99 14 90	83 26 03	15 00 71
43	75 53 11	01 11 11	78 56 62	03 87 34	18 12 42
44	09 30 87	33 32 37	96 79 07	33 75 21	74 06 47
45	02 30 44	66 34 64	38 75 01	40 22 87	76 19 01
46	14 45 74	30 52 97	77 13 20	66 87 54	89 05 30
47	82 45 49	85 02 33	58 84 03	74 63 52	15 47 04
48	44 33 94	98 75 51	62 00 17	59 00 42	09 39 66
49	96 00 26	82 60 22	02 60 69	99 09 67	01 12 01
50	20 67 56	12 77 16	78 04 36	38 95 35	71 26 49
51	64 60 21	12 41 60	04 63 93	45 25 52	75 50 35
52	64 76 41	17 07 54	01 29 86	41 93 16	55 54 40
53	93 46 82	67 64 48	91 74 85	94 40 51	30 93 08
54	82 64 44	58 45 94	30 39 86	19 64 84	35 30 19
55	61 46 40	89 21 47	20 85 91	90 56 67	40 31 46
56	92 80 33	89 23 96	24 33 16	80 45 20	35 36 00
57	45 65 20	02 56 40	21 35 17	71 33 07	36 71 90
58	40 99 02	66 37 59	24 79 35	21 61 29	96 50 01
59	50 31 47	84 44 30	70 33 12	63 54 86	63 08 62
60	87 38 68	91 50 98	65 95 29	54 20 89	25 59 33
61	20 85 20	04 05 21	53 58 65	40 60 53	91 07 32
62	03 16 83	56 90 30	18 77 83	18 91 26	70 50 99
63	57 12 84	22 89 61	19 55 62	96 01 44	28 96 21
64	11 06 38	37 58 65	66 54 73	80 38 57	44 99 81
65	44 68 39	54 96 66	56 83 21	22 34 00	28 65 20
66	73 19 05	41 32 92	36 98 10	94 60 47	32 10 82
67	39 56 14	02 45 65	16 86 78	90 46 39	58 62 66
68	32 53 16	30 76 36	80 52 65	02 10 07	81 40 80
69	98 43 67	05 82 06	19 24 86	24 30 44	06 15 54
70	53 08 00	94 46 80	60 94 01	83 94 45	42 43 55
71	28 21 05	43 60 40	73 70 75	33 10 74	91 83 95
72	89 79 63	50 98 53	56 42 12	76 48 56	34 46 82
73	61 48 17	25 59 95	19 14 31	68 94 23	83 40 83
74	41 98 20	72 70 69	39 46 17	37 70 37	81 75 23
75	51 08 35	35 16 20	92 94 25	05 04 01	65 33 82
76	73 97 76	94 92 07	24 89 41	98 35 91	96 52 82
77	43 74 49	01 59 38	60 29 94	61 02 11	61 86 36
78	94 94 39	87 49 44	54 02 52	56 28 49	34 49 25
79	52 10 65	11 34 68	68 65 58	90 17 33	98 36 82
80	54 42 73	62 51 54	80 63 36	65 12 44	52 16 12
81	73 27 51	94 71 14	37 55 00	05 32 36	59 89 86
82	77 69 59	62 33 99	26 67 95	72 77 16	02 28 96
83	08 19 98	26 68 06	02 05 57	21 73 55	35 07 79
84	50 83 92	60 44 28	52 83 25	39 83 60	92 71 10
85	16 89 30	82 48 70	63 82 71	48 72 82	77 37 56
86	21 41 74	65 08 73	82 94 72	22 67 92	34 74 33
87	99 08 47	77 43 94	17 07 76	57 93 68	61 15 97
88	20 02 69	70 87 44	57 23 35	99 94 16	63 40 99
89	93 95 15	81 21 75	71 39 23	31 06 43	87 44 21
90	10 91 65	40 88 43	50 57 83	50 82 34	12 78 80
91	91 72 35	36 80 19	49 49 37	17 40 98	02 53 59
92	23 82 82	20 56 34	76 49 27	40 78 29	99 07 22
93	21 02 08	25 07 15	36 45 19	21 30 48	30 76 99
94	80 22 31	36 25 82	63 91 94	56 59 42	34 18 65
95	46 92 44	39 46 22	03 99 15	60 45 34	06 77 86
96	54 81 74	93 71 51	14 28 22	66 21 53	31 66 09
97	80 17 11	33 37 07	00 77 89	31 86 72	35 67 41
98	05 47 12	99 05 06	18 52 83	53 36 90	84 63 41
99	78 99 91	58 03 59	93 60 31	40 23 58	35 51 66
100	20 33 54	25 07 06	55 95 53	14 64 58	01 07 63

Solutions to activities

Solution to Activity 1

The labels selected are

52 10 65 11 34 68 58 90 17 33 98 36.

The list cannot be obtained by simply taking the first 12 pairs along the row; the seventh pair is 68, which has already appeared in the sample, and the eighth pair is 65 which has also already appeared, so you should have ignored the seventh and eighth pairs.

Solution to Activity 2

The labels selected are

72 22 05 84 39 89 57 73 84 86 57 76 79 08 65
47 14 97 61 57 30 93 88 12 88 58 15 75 17 45.

This time there is no problem with repeated individuals in the sample.

Solution to Activity 3

The sample is shown in the following table:

Name	Label	Gender	Occupation
Hare, Dorothy	41	F	P
Dev, Mohen	25	M	P
Redman, Guy	66	M	P
Crofts, Mary	21	F	A
Lang, Chris	51	M	P
Bramley, Max	11	M	P
Graham, Bert	34	M	P
Gowan, Dai	33	M	P
Cluskie, Alex	18	M	P
Grant, Lynne	36	F	P
Rowan, George	70	M	P
Ricardo, Dan	68	M	P
Masterton, Dick	58	M	P
Sandford, Dave	71	M	P
Damper, Emma	24	F	S
Bates, Sheila	06	F	S
Woodhouse, Paul	84	M	M
James, Patricia	48	F	A
Franks, Abraham	32	M	P
Fallow, Jim	29	M	P

The sample of size 20 that you have just obtained is rather more representative of the population than was the previous sample of size 10. In this larger sample, 70% are men and 30% women, compared to 60% and 40% in the population. In addition, this sample fairly closely represents the occupational pattern in the population. It slightly over-represents the professional staff and under-represents secretarial staff. In a sample of size 20, you might expect about four secretarial staff; this sample has only two. However, this larger sample should represent the population quite well for most practical purposes. (That is not to say, of course, that every simple random sample of size 20 would represent the population as well!)

Solution to Activity 4

Step 1 The first pair of digits from row **3** in the range 01 to 17 is 06, so this is the random start. (Notice that you must use pairs of digits; you cannot use the single digit 6 at the beginning of the line.)

Step 2 The labels in the sample are every 17th label:

06 23 40 57 74.

The sample is shown below.

Name	Label	Gender	Occupation
Bates, Sheila	06	F	S
Daley, Stuart	23	M	P
Hallow, Jean	40	F	A
McCraig, Frank	57	M	P
Stratford, Peter	74	M	P

For such a small sample, this is about as representative of the target population as you might hope. There are three men and two women, which is the same ratio as in the population. Also, there are three professionals, one member of the secretarial staff and one administrator; this is a fair representation of three of the categories. There are no manual workers in this particular sample.

Solution to Activity 5

Step 1 The first digit in row **29** is 3, so we start at label 03.

Step 2 The labels in the sample are every fourth label:

03 07 11 15 19 23 27 31 35 39 43 47 51 55 59 63 67 71 75 79 83.

The sample is shown in the following table.

Name	Label	Gender	Occupation
Archer, Simon	03	M	M
Baxter, John	07	M	P
Bramley, Max	11	M	P
Chapman, Liz	15	F	M
Cramer, Will	19	M	P
Daley, Stuart	23	M	P
Eric, Steve	27	M	P
Foster, Sue	31	F	S
Graham, Bill	35	M	P
Greenway, Maggie	39	F	P
Hewitt, Ray	43	M	P
Iron, Donald	47	M	P
Lang, Chris	51	M	P
Lupton, David	55	M	P
Menton, Christine	59	F	S
Osterley, Rebecca	63	F	S
Redstar, Pamela	67	F	S
Sandford, Dave	71	M	P
Thompson, Anna	75	F	S
Turner, Richard	79	M	P
Winston, Chuck	83	M	P

There are 14 men in the sample of 21, which is 67% compared to 59% of the target population. There are also 14 professionals (67%) compared to 65% in the target population. 24% of the sample are secretarial staff, compared with 21% of the population. There are two manual workers but no administrators. On the whole, this sample provides quite a good representation of the target population. The lack of representativeness is not really any more than one might expect in a sample of this size.

Solution to Activity 6

Since the batch size is 1000, the median is halfway between the 500th and 501st values. Counting in 500 from the 'Much worse off' end of the population, responses 1 and 2 ('Much worse off' and 'Somewhat worse off') account for 400 values, so the 500th value is 3. Similarly the 501st value is also 3, so the median is 3.

Solution to Activity 7

If we put the responses in each batch in ascending order, then the median of each is the middle value as given below. (Obviously you could determine the middle value of three numbers without writing them down.)

Sample	Ordered responses			Median
A	1	2	4	2
B	1	4	5	4
C	1	4	4	4
D	1	3	5	3
E	1	1	1	1
F	3	5	5	5

Solution to Activity 8

The most noticeable, and most important, change is that, as the sample size increases, the proportion of samples with median 3 increases, whilst the proportions with medians 1, 2, 4 and 5 decrease.

For $n = 15$, already over half of the samples (actually about 0.574 of them) have median 3, and for $n = 45$ this proportion has risen even higher, to 0.827. For $n = 135$, nearly all the samples (a proportion of 0.981) have median 3.

Solution to Activity 9

The value with the largest proportion is the one with the longest vertical bar. This value is 4. (The proportion of the samples with median response 4 is 0.324.)

Solution to Activity 10

For each sample size n pictured, the proportion of the samples of size n with median 3 is as follows:

n	Proportion
3	0.296
5	0.365
9	0.467
15	0.574
25	0.692
45	0.827
75	0.921
135	0.981

Thus those sample sizes for which this proportion is larger than 60% (i.e. 0.6) are 25, 45, 75 and 135.

Solution to Activity 11

To choose strata, you need information that is both related to the subject under investigation and available for all individuals in the population before the survey starts. The only information that is mentioned as being available for all customers is name, address and quantity of seed bought. A customer's address is likely to be related to the geographical location where the customer grew the seeds, and satisfaction with the results might well be related to location because climate varies with location. Therefore, it would make sense to stratify in terms of geographical region. You might also have felt that a customer's satisfaction might be related to the amount of seed bought; if so, that could also be used for stratification.

You may have suggested other criteria for stratification, and these may well be sensible, but remember that a variable used for stratification needs to be known for all the customers before the sample is chosen.

Solution to Activity 12

The random numbers in row **6** start 71 58 45 .... Hence we start with the 7th person listed in Table 11, Mary Abbott. She is the first person in the sample and we then include every eighth person until we reach the end of the list. From the table, the people in the sample are: Mary Abbott (Y), Arthur Mason (N), Jacqueline Groves (Y), Sean Daly (Y), Mary Dunn (Y), Raymond McCann (Y), Nicholas Kohler (N), Annette Clarke (N), Jaya Shah (Y), Darrell Paine (N), Dipak Shah (Y) and Muriel Uttley (Y). In this sample of 12, the number of bus users is eight, so the sample estimate of the percentage of bus users in the population is again $8/12 \simeq 66.7\%$.

Solution to Activity 13

The first two single digits in row **26** are 3 and 2, which correspond to Lytham Gardens and Kerrera Close.

Lytham Gardens has 12 electors, so we will select $12/3 = 4$ of these. Kerrera Close has 27 electors so we will select $27/3 = 9$ of these.

Starting in row **82**, the random number pairs are as follows. (The pairs corresponding to selected registration numbers are given in italics.)

82	77 69 59	62 *33* 99	26 67 95	72 77 16	02 *28* 96
83	08 19 98	26 68 06	02 05 57	21 73 55	*35* 07 79
84	50 83 92	60 44 28	52 83 25	*39* 83 60	92 71 *10*
85	*16* 89 30	82 48 70	63 82 71	48 72 82	77 37 56
86	*21* 41 74	65 *08* 73	82 94 72	*22* 67 92	34 74 33
87	99 08 47	77 43 94	*17* *07* 76	57 93 68	61 *15* 97
88	*20* 02 69	...			

The selected registration numbers for Lytham Gardens (from registration numbers 1028–1039) are: (10)33 (10)28 (10)35 (10)39.

Those for Kerrera Close (from registration numbers 1001–1027) are: (10)10 (10)16 (10)21 (10)08 (10)22 (10)17 (10)07 (10)15 (10)20.

Thus the people in the survey and their bus usages are: Nicholas Kohler (N), John Kerr (N), Marie Clements (N), Samuel Bolton (N), Audrey Ho (N), Garry Dunn (Y), Matthew Goulding (Y), Mary Dixon (Y), Janet Goulding (Y), Mary Dunn (Y), Alan Smith (Y), Clare Christon (Y) and Eleanor Price (N).

In this sample of 13, the number of bus users is seven, so the sample estimate of the percentage of bus users in the population is $7/13 \simeq 53.8\%$.

Solutions to exercises

Solution to Exercise 1

There are many ways of using the table to choose such a sample. Perhaps the most straightforward method uses groups of three digits, working along the rows from a randomly chosen starting point much as you did for the other two target populations in Subsection 1.2.

For example, if the starting point is the beginning of row **49**, then this method will select the following labels:

960 026 826 022 069 990 967.

With this starting point, the individual 026 was repeated and had to be ignored the second time. There may have been a problem with repeated individuals in your sample, but this is quite unlikely with a small sample from a large population.

Solution to Exercise 2

(a) The nine labels selected are

26 25 37 27 02 15 51 87 18.

To obtain this sample it is necessary to use 11 digit pairs from the table, because the labels 26 and 27 are repeated.

(b) The 17 labels selected are

32 57 79 72 02 27 96 10 62 63 07 30 01 40 97 69 23.

This time there is no problem with repetition: 17 digit pairs are enough.

Solution to Exercise 3

(a) The first eight pairs of digits from row **5** in the range 01 to 86 are

50 76 11 36 13 84 32 72.

The following two tables show the people in this sample and analyse the sample by gender and occupation.

Name	Label	Gender	Occupation
Kapoor, Sashi	50	M	P
Thompson, Jack	76	M	P
Bramley, Max	11	M	P
Grant, Lynne	36	F	P
Cameron, Lynne	13	F	P
Woodhouse, Paul	84	M	M
Franks, Abraham	32	M	P
Shah, Anjali	72	F	S

	Male	Female	Total
Professional	4	2	6
Administrative	0	0	0
Secretarial	0	1	1
Manual	1	0	1
Total	5	3	8

(b) The sample and its analysis are shown in the following tables.

Name	Label	Gender	Occupation
Singh, Meera	73	F	S
Bidford, David	09	M	P
Archer, Simon	03	M	M
London, Fred	54	M	P
Crofts, Dennis	20	M	P
Andrews, Jean	02	F	P
Lupton, David	55	M	P
Jolly, Susan	49	F	S
James, Patricia	48	F	A
Hutton, Joan	46	F	S
Thompson, Anna	75	F	S
Harrison, Sheila	42	F	P

	Male	Female	Total
Professional	4	2	6
Administrative	0	1	1
Secretarial	0	4	4
Manual	1	0	1
Total	5	7	12

(c) We must select every ninth label starting at label 05. Hence the sample is as follows.

Name	Label	Gender	Occupation
Baker, Fred	05	M	P
Carter, Jane	14	F	P
Daley, Stuart	23	M	P
Franks, Abraham	32	M	P
Hare, Dorothy	41	F	P
Kapoor, Sashi	50	M	P
Menton, Christine	59	F	S
Ricardo, Dan	68	M	P
Trumpington, Pat	77	F	S
Yeo, Tara	86	F	A

The following is an analysis of the sample.

	Male	Female	Total
Professional	5	2	7
Administrative	0	1	1
Secretarial	0	2	2
Manual	0	0	0
Total	5	5	10

(d) This time we must select every tenth label starting at label 08, giving the following sample.

Name	Label	Gender	Occupation
Best, John	08	M	P
Cluskie, Alex	18	M	P
Estover, Matthew	28	M	P
Greenson, Denise	38	F	A
James, Patricia	48	F	A
Masterton, Dick	58	M	P
Ricardo, Dan	68	M	P
Truscott, Karen	78	F	S

The following is an analysis of the sample.

	Male	Female	Total
Professional	5	0	5
Administrative	0	2	2
Secretarial	0	1	1
Manual	0	0	0
Total	5	3	8

Solution to Exercise 4

(a) 0.099.

(b) 0.168.

(c) To have a median less than 3 the sample must have median 1 or 2. So the proportion of samples with median less than 3 is the sum of the proportions with medians 1 and 2. This is $0.099 + 0.168$, which equals 0.267.

(d) Similar reasoning implies that this is the sum of the proportions of samples with medians 4 and 5. This is $0.266 + 0.001 = 0.267$.

Note that the following proportions sum to one, approximately. The digit 1 in the last decimal place is due to rounding in the calculations.

Proportion with median less than 3	0.267
Proportion with median 3	0.467
Proportion with median greater than 3	0.267
Sum	1.001

The sum would be expected to be equal to 1 because each sample median is either less than 3, equal to 3 or greater than 3.

Solution to Exercise 5

(a) The population size is 1000, so the median is halfway between the 500th and 501st values. Counting in 500 from the 'Much worse off' end of the population, the 500th and 501st values both equal 2. Hence the median is 2.

(b) In Figure A, the proportion of samples that give a median of 2 is very small. As the population in the table comes from a population with a median of 2, Figure A must be the sample that relates to a different population. Looking at Figures B and C, the median is far more predictable from Figure B than from Figure C, so Figure B must relate to the larger sample. Thus Figure B is for a sample of size 21 from the tabulated population, while Figure C is for the sample of size 7.

Solution to Exercise 6

(a) Cluster sampling would be valid and appropriate, because the expenditure pattern of such households is unlikely to be related to geographical area.

(b) It would be difficult to obtain a valid sampling frame as there is no simple way to identify which households own their home outright and which do not. Therefore quota sampling would have to be used.

Solution to Exercise 7

(a) Two lists could be obtained from official records: all addresses, and all addresses with a registered motor vehicle. From this, a list of all addresses at which no vehicles are registered could be obtained.

(b) Either a simple or a systematic random sample would be sufficient, particularly as this is just a pilot survey.

Solution to Exercise 8

(a) The sizes of the three strata are Jersey Close: 22; Kerrera Close: 27; other three roads: 46, which together total $22 + 27 + 46 = 95$. A total sample of size 12 is required, so the numbers to take from each stratum are:

$$\text{Jersey: } \frac{22}{95} \times 12 \simeq 3, \quad \text{Kerrera: } \frac{27}{95} \times 12 \simeq 3, \quad \text{other: } \frac{46}{95} \times 12 \simeq 6.$$

These sample sizes add to 12.

(b) Starting at the beginning of row **52**, the selected registration numbers for Jersey Close (977–1000) are: 986 993 982.

For Kerrera Close (1001–1027): 1008 1019 1021.

For the third stratum (1028–1074): 1047 1056 1067 1040 1031 1046.

Hence the electors in the sample and their bus usages are: Margaret Donegan (N), Joan Mason (Y), Linda Jones (Y), Mary Dixon (Y), Vince Edwards (Y), Matthew Goulding (Y), Brian Ferguson (N), David McGregor (N), Mala Shah (N), David Clarke (N), David Kerr (Y) and Angela Smith (Y).

(c) Seven individuals in this sample of 12 people are bus users, so the sample estimate of the percentage of bus users in the population is

$$\frac{7}{12} \times 100\% \simeq 58.3\%.$$

In the target population of 95 electors, there are 49 people who used the bus in the previous week, so the true percentage of bus users is $49/95 \simeq 51.6\%$. Hence the sample estimate is reasonably close to the population value. (For a sample of 12, the only sample result that would be closer is when the sample contains six bus users, which is only one different from the number in the sample we selected.)

Acknowledgements

Grateful acknowledgement is made to the following sources:

Cover image: Minxlj/www.flickr.com/photos/minxlj/422472167/. This file is licensed under the Creative Commons Attribution-Non commercial-No Derivatives Licence http://creativecommons.org/licenses/by-nc-nd/3.0/

Introduction, cartoon (Tower of Pisa), www.causeweb.org

Figure 2 Taken from:
www.ons.gov.uk/ons/guide-method/census/2011/index.html and http://www.scotlandscensus.gov.uk/en/

Figure 3 © 2012 Microsoft Corporation

Figure 4 Taken from: Google Images

Figure 6 © 2012 Microsoft Corporation

Figure 11 David Ayres

Figure 12 This file is licensed under the Creative Commons Attribution Licence http://creativecommons.org/licenses/by/3.0/

Subsection 1.2 figure, 'A UK National Lottery machine', taken from: http://www.mirror.co.uk/money/city-news/lottery-set-for-42billion-boost-as-operator-753620

Subsection 3.1 cartoon (hard to quantify), LightBulb Cartoon

Subsection 4.1 photo of George Gallup: Gallup Inc.

Subsection 4.1 cartoon (margin of error): John Landers, www.causeweb.org

Every effort has been made to contact copyright holders. If any have been inadvertently overlooked the publishers will be pleased to make the necessary arrangements at the first opportunity.

Unit 5

Relationships

Introduction

In Units 2 and 3, we looked at prices and incomes and attempted to answer the question: *Are people getting better or worse off?* You have learned several statistical techniques for summarising a batch of data and for comparing two batches. In Unit 4, you saw how to choose a sample for a survey and you were introduced to ideas about random sampling.

In this unit, we are not going to attempt to answer any particular question, but we are going to investigate relationships between two variables. Scatterplots can be used to picture such relationships, and they were used for this purpose in Unit 1. For instance, in Subsection 2.1 of Unit 1 some data on the quantity of fertiliser and yield of wheat-grain produced were given in a table. A scatterplot was then used to explore the relationship between the quantity of fertiliser applied and the yield of grain. The scatterplot is reproduced in Figure 1. It shows that grain yield increased as more fertiliser was applied. Moreover, the plotted points lie roughly in a straight line, suggesting that a straight line could be used to model the relationship between the two variables. This happens quite often when two variables are related – an aim of this unit is to give a way of calculating a straight line that best represents the relationship between the variables.

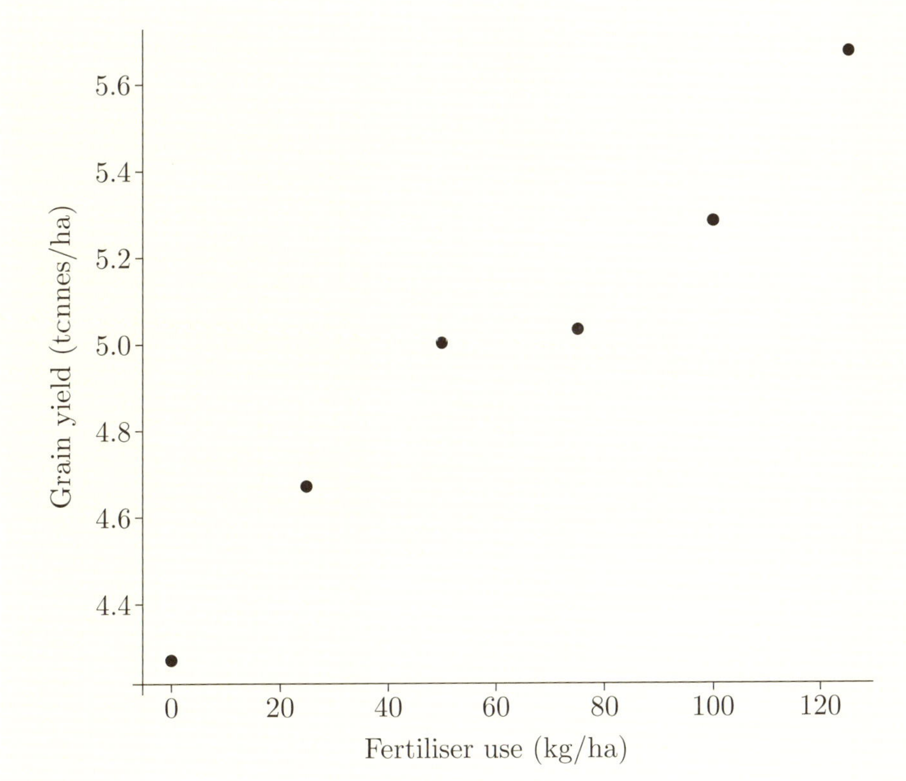

Figure 1 Grain yield by fertiliser use

We also explore other information that scatterplots yield both about relationships and about individual data points, paying particular attention to unusual points that are a long way from the main body of data. After drawing a straight line on a scatterplot, we address the question of how to tell when the line is a good way of representing the relationship between the two variables.

Section 1 describes what is meant by a *relationship* between two variables; it introduces scatterplots in detail, in particular addressing which variable should be plotted on which axis. In Section 2, scatterplots are used to characterise different types of relationship and identify unusual data points. Drawing an appropriate line on a scatterplot to represent a relationship is first considered in Section 3, where the line is drawn 'by eye'. Calculating the *least squares fit line* to model the relationship is the topic of Section 4. Uses to which this line can be put are described in Section 5.

Section 6 directs you to the Computer Book. You are also guided to the Computer Book at the end of Subsection 4.1. This work is critical to the unit and it is strongly recommended that you do it at this point in the text. However, you will be reminded in Section 6 in case you have not completed it by then.

1 Relationships and scatterplots

Statisticians refer to quantities like age, height or price, which vary from one individual or one purchase to another, as **variables**. Two variables are said to be **related** if knowing the value of one of the variables provides information about the value of the other variable. In this unit we shall be looking at various questions that can be asked about related variables.

- How can we investigate whether two variables are related?
- How can we describe a relationship between two variables numerically?
- What use can we make of a numerical description of a relationship?
- How can we interpret or explain a relationship?

First, though, we consider more closely what constitutes a relationship between two variables.

1.1 What is a relationship?

Suppose you are the membership secretary of a sports club for which the minimum age of entry is 10 years. You have to apply this rule, but obviously you do not want to upset potential members. One day two girls arrive and say they would like to join. One is about ten centimetres taller than the other. You ask the taller girl how old she is and she replies that she is 10. So you guess that the other girl will be too young to join; however, when asked, she says that she is 12. You are surprised because you based your guess on the fact that taller girls are usually older. In other words, there is a **relationship** between age and height of girls. It is not a perfect relationship because a 12-year-old girl may be shorter than a 10-year-old, and, assuming the two girls are telling the truth, this is what happened in the case above.

Suppose your son offered to do the weekly shopping at the supermarket and you asked him to get 10 kg of potatoes, although usually you only buy 5 kg of potatoes. You would not know exactly how much these would cost, because the price varies between varieties and from week to week. Also, a bag containing 10 kg of potatoes usually costs a little less than two 5 kg bags. However, you would probably have some idea of how much these would cost, as the weight of potatoes provides a guide to this.

These two situations both involve relationships between two variables. In the case of the sports club, knowing the girls' heights enabled you to guess at the girls' ages (wrongly, as it turned out). The relationship applies both ways; knowing a child's age would give you information about his or her height. It is not precise information. For example, if you were told that a girl was eight years old today, you could not say that she was exactly 1.25 metres tall. However, you could be fairly certain that she would be shorter than a 12-year-old girl, and you could say (given the appropriate information) that she would probably be between 1.18 and 1.32 metres tall.

In the potato example, you probably thought in terms of price per kilogram. However, price is just a way of describing the relationship between weight of potatoes and amount of money paid. Generally, the more you buy, the more you pay.

Activity 1 *Investigating height and age*

To start you thinking about what is involved in learning about a relationship, try answering some questions related to the sports club situation above.

(a) How would you investigate the relationship between height and age in children?

(b) How might you describe the numerical relationship between height measurements and age values of children?

(c) Would you expect to see the same sort of relationship between age and height in adults as you would see in children?

1.2 Linked data

Let us now turn to a different example of a relationship – that between car ownership and unemployment. That is, if we know the rate of unemployment in an area, does that tell us anything about the amount of car ownership in the area?

Example 1 *Unemployment in Bedfordshire; car ownership in Merseyside*

In the UK ten-yearly census, all households are required to complete a detailed return, and this provides information on many topics, including car ownership and unemployment. Each household records (amongst many other things) the number of cars owned by members of the household, the number of men aged 16–74 in the household, and the number of those men who are unemployed on the date of the census. A 10% sample of these data is analysed and the results are published in the form of percentages for every town and region in the UK. Tables 1 and 2 show some of the results from the 2001 census. Table 1 shows the percentage of men unemployed in four regions of Bedfordshire. Table 2 shows the percentage of households with no car in five regions of Merseyside.

Table 1 Male unemployment in Bedfordshire

Bedfordshire	Percentage of men unemployed
Bedford	3.99
Luton	4.82
Mid Bedfordshire	2.07
South Bedfordshire	2.73

(Source: HMSO (2004) *Census 2001: Key Statistics for Local Authorities in England and Wales*, Table KS09b)

Table 2 Access to cars in Merseyside

Merseyside	Percentages of households with no car
Knowsley	41.76
Liverpool	48.28
St Helens	30.48
Sefton	31.00
Wirral	30.34

(Source: HMSO (2004) *Census 2001: Key Statistics for Local Authorities in England and Wales*, Table KS17)

Activity 2 *Considering unemployment and car ownership*

Do the figures in Tables 1 and 2 above provide any information about a possible relationship between household car ownership and male unemployment rates?

As you saw in Activity 2, to investigate the relationship between car ownership and unemployment, we need linked data giving both percentages for a number of towns. For convenience, 'town' means town or small region for the remainder of this example.

Linked data

Data are said to be **linked** when two or more variables are recorded for the same sampling units.

When there are two variables, linked data are also often referred to as *paired data.*

Data from the UK Census in 2001 includes unemployment rates and rates of car ownership for towns in Great Britain. Because of the large number of towns in Great Britain, a stratified sampling scheme was used to select the data below. The sampling was limited to England and used the main regions as strata. One town or small region was selected from each of West Midlands, North West, Yorkshire and the Humber, North East, East Midlands, South West, and East, and three towns were selected from the South East region, which is the most populated. London and the major cities were omitted because they might not be typical of the country as a whole. Both percentages (the 'variables' in this case) were recorded for each of the ten towns (the 'sampling units'); the linked data are shown in Table 3.

Table 3 Male unemployment and car ownership for ten towns in England

Town	% males unemployed	% households with no car
Alnwick, North East	4.59	21.6
Vale Royal, North West	3.55	17.2
Rotherham, Yorkshire and the Humber	5.19	29.7
Rutland, East Midlands	1.75	13.6
Dudley, West Midlands	5.27	25.3
Norwich, East	5.61	35.5
Bracknell Forest, South East	2.25	14.5
Rother, South East	3.00	20.8
Mole Valley, South East	1.84	13.1
West Dorset, South West	2.14	16.9

(Source: HMSO (2004) *Census 2001: Key Statistics for Local Authorities in England and Wales*, Tables KS09b and KS17)

In Table 3 each row gives the percentage of males unemployed for that town and the percentage of households with no car for the *same* town. So you can get some idea of the relationship between male unemployment and car ownership by just looking at the two columns of numbers. For example, Rutland and Mole Valley have the lowest percentages in both columns, whereas Rotherham and Norwich both have a high percentage of male unemployment and households with no car.

Note that 'relationship' is not meant to imply anything about *causality*. That is, a relationship between male unemployment and lack of cars does not imply that unemployment causes lack of car ownership, or that a lack of cars causes unemployment.

Activity 3 *Linked or not linked?*

For each of the two sets of data described below, state whether the data are linked data or not.

(a) Measurements of heights of two groups of children: one group of twenty year-6 children and one group of twenty year-7 children.

(b) Measurements of height for one group of twenty year-7 children, both one year ago and now.

1.3 Scatterplots

In the previous subsection you saw that linked data can be displayed in a table. Relationships in the data can be explored by looking in the table for patterns. However, a scatterplot gives us a better impression of linked data, making it easier to spot patterns. Figure 2 shows a scatterplot of the unemployment and car ownership data given in Table 3.

Alternative expressions for scatterplot are **scattergram**, **scattergraph** and **scatter diagram**.

Figure 2 A scatterplot of car ownership against unemployment

In Figure 2, the horizontal axis (the x-axis) is labelled 'Percentage of men unemployed', and the vertical axis (the y-axis) is labelled 'Percentage of households with no car'. These axes represent the two columns of data we are displaying in the plot. The scales have been chosen to cover the range of the data (from 1.75 to 5.61 for the percentage of males unemployed, and from 13.1 to 35.5 for the percentage of households with no car). The numbering of scales is done with numbers that can be written down using just a few significant figures. As happens in many scatterplots, the plotted scales do not start at zero. Instead the range of values plotted for each axis is chosen so that points on the scatterplot cover as much of the area of the scatterplot as possible.

Computer software for drawing scatterplots is usually able to choose the scales automatically.

There are ten points marked on the scatterplot. These points represent the ten towns in Table 3. The position of each town's point is given by the two values in the corresponding row on Table 3. For example, in Dudley at the time of the census in 2001, 5.27% of men were unemployed and 25.3% of households had no car. So the point representing Dudley is placed at the position corresponding to 5.27 along the horizontal axis and 25.3 along the vertical axis. The position of points on a scatterplot can be written concisely using **coordinates**. For example, the values 5.27 and 25.3 are the coordinates of the point representing Dudley: 5.27 is called the **first coordinate** or **x-coordinate**, and 25.3 is called the **second coordinate** or **y-coordinate**.

A wide variety of symbols can be used to mark points on a scatterplot, including dots or crosses.

In order to emphasise that the values 5.27 and 25.3 are the coordinates for a point on the scatterplot, it is common to write the two values side by side, separated by a comma and enclosed in brackets like this:

$$(5.27, 25.3).$$

The first number in the bracket is the value along the horizontal axis and the second number is the value along the vertical axis. It is important always to write the numbers in this order in the brackets. The coordinates (25.3, 5.27) would tell us that for some town in 2001, 25.3% of men were unemployed and that 5.27% of households had no car.

One way of remembering which way round to write the numbers in brackets is that in the alphabet 'h' comes before 'v', and so the coordinate along the *h*orizontal axis comes before the coordinate on the *v*ertical axis.

Activity 4 *Exploring a scatterplot*

(a) Using the data in Table 3 write down the coordinates for the following two towns: Vale Royal and Rother.

(b) Which town is represented by the point in the top rightmost corner of Figure 2?

(c) Describe in words what the scatterplot tells you about the relationship in this batch of data.

Can you see the upper points of my scatterplot?

Activity 4 asked to you describe the scatterplot in Figure 2. The description and interpretation of scatterplots will be considered in more detail in Section 2.

1.4 Response and explanatory variables

One important aspect of constructing a scatterplot has not yet been mentioned: which variable to put on the x-axis and which variable to put on the y-axis. When investigating the relationship between two variables it often happens that the values taken by one variable can be partly explained using the values taken by the other variable. For example, the height of a child can be partly explained by the child's age. In such situations, the variable being explained (such as child's height) is called the *response* variable, and the variable doing the explaining (such as child's age) is the *explanatory* variable.

Sometimes it makes more sense to think that the value of one variable depends, at least in part, on the value of the other variable. In this case the response variable is the variable that depends on ('responds to') the other variable, and the variable on which the response variable depends is the explanatory variable. To illustrate, suppose you are given some linked data on petrol consumption of a car: various speeds and the miles per gallon when the car travelled at each speed. In this case, miles per gallon is the response variable, as it depends, to a certain extent, on the speed of the car, which is the explanatory variable.

Finally, sometimes the value of one variable is to be predicted, and the value it takes is partly related to the value of the second variable. For example, you might want to predict the miles per gallon you will obtain if you drive at 65 mph. Then the variable to be predicted is the response variable and the second variable is the explanatory variable.

Explanatory and response variables

A **response variable** is the variable that is being explained or whose value depends on other variables. It is also the variable to be predicted if predictions are to be made. It is sometimes known as the *dependent* variable.

An **explanatory variable** is the variable that is doing the explaining or is the variable on which the response variable depends. It is sometimes known as the *independent* variable.

By convention, on a scatterplot the explanatory variable is put on the x-axis and the response variable is put on the y-axis.

Example 2 *Blood pressure*

Suppose that you are given some linked data on blood pressure that consists of blood pressure measurements on patients before and after a treatment. In this case, the blood pressure measurement after treatment is the response variable, as it depends to a certain extent on the blood pressure before treatment, which is the explanatory variable. It does not make sense to think that someone's blood pressure before the treatment can be changed by changing their blood pressure after the treatment.

So, on a scatterplot the blood pressure before the treatment would be plotted along the x-axis, and the blood pressure after the treatment would be plotted along the y-axis.

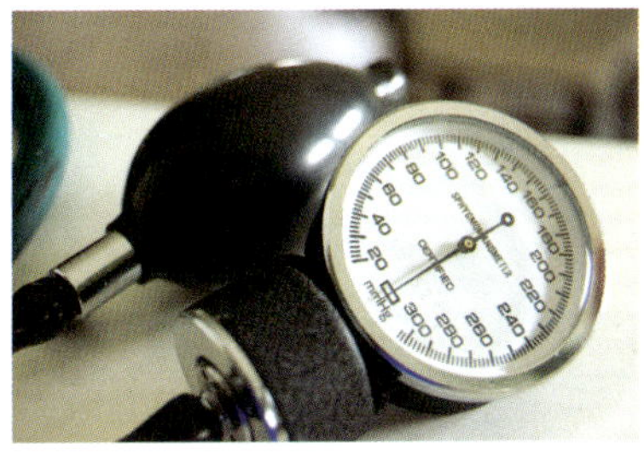

A sphygmomanometer, an instrument used to measure blood pressure

Example 3 *Household expenditure*

Suppose in a survey of household expenditure, 12 households are asked to record their total expenditure for one week and also to note what items they bought.

If a household has a low income, it spends this on necessities, including food, and cannot afford luxuries. However, when income increases, more money will be spent on luxuries. Although the household will probably spend more on food, maybe paying for higher quality, the increase is proportionately less, and so the percentage of total income spent on food falls.

So when we draw a scatterplot of total expenditure and percentage of total income spent on food, total expenditure is the explanatory variable and percentage spent on food is the response variable. This means that total expenditure would be put on the x-axis, and the percentage of total expenditure spent on food would be put on the y-axis. This makes sense as a household is very unlikely to decide what percentage of its total expenditure should go on food before working out what its total expenditure should be.

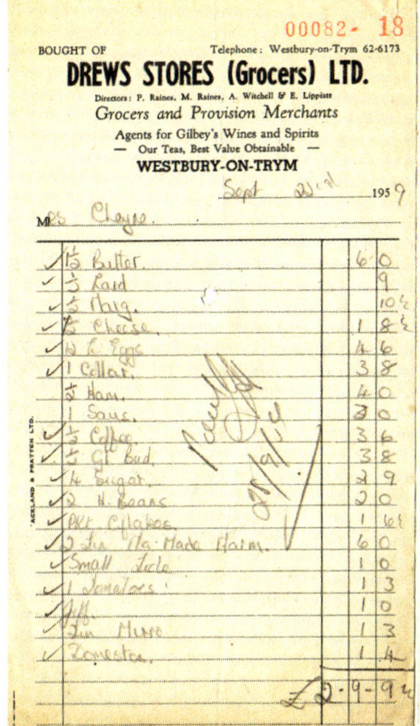

Some old household expenses

In most experimental situations, it is usually clear which is the explanatory variable. An experiment often consists of choosing values of an explanatory variable (for example, amount of fertiliser applied, dose of a drug given to patients, temperature of an industrial process) and then observing the effect on the response variable (for example, yield of tomatoes, blood pressure of patients, strength of a manufactured component).

Sometimes, though, the use that will be made of the data determines which variable is the response variable and which is the explanatory variable. If we wish to forecast one variable when the other takes a particular value, the variable we wish to *forecast* is regarded as the response. For example, if a married man's height is to be predicted from the height of his wife, then the man's height is the response and the wife's height is the explanatory variable. These roles are reversed if a married woman's height is to be predicted from the height of her husband. If the use of the data is unspecified, then it is arbitrary which of their heights should be plotted on the x-axis and which on the y-axis.

Activity 5 *Which variable would you plot on the x-axis?*

For each of the following cases, if you were asked to draw a scatterplot of the data, which variable would you choose as the x-coordinate? Give a reason for your choice.

(a) In order to investigate the effects of different amounts of fertiliser on the yield of tomatoes, ten tomato plants of the same variety were each given a different amount of the same fertiliser. The data consist of the amount of fertiliser and the weight of tomatoes for each plant.

(b) The data consist of the numbers shown in Table 3 (Subsection 1.2): the percentage of males unemployed and the percentage of households with no car in a random sample of ten towns in England.

(c) For a series of water companies in the UK, the average consumption by households with water meters and the average consumption by households without water meters were collected.

Some tomatoes growing on a plant

Exercises on Section 1

Exercise 1 *Linked or not?*

For each of the following pairs of variables, state whether you think they are linked or not. Justify your opinion.

(a) The heights of a group of twenty 5-year-old children in one school and the weights of twenty 5-year old children in a different school.

(b) The heights of twenty 5-year old children and weights of another twenty 5-year-old children, all from the same school.

(c) The heights and weights of twenty 5-year-old children.

A depiction of some children

Exercise 2 *Identifying explanatory and response variables*

For the following pairs of linked variables, discuss which variable could be regarded as the response variable and which as the explanatory variable.

(a) Average house price and calendar year.

(b) Average hourly wage earned by men and average hourly wage earned by women, in different sectors of the economy.

(c) In a study to predict employment rates, the unemployment rate in different countries and the employment rate in those countries.

2 Interpreting scatterplots

In Section 1, the use of linked data to investigate relationships in data was introduced. You also saw that such data can be displayed graphically as a scatterplot. In this section, we shall investigate what can be learned from looking at a scatterplot.

When interpreting a scatterplot, we are only concerned with a general overall relationship. That is, the general pattern set by the vast majority, if not all, of the points. Any points that do not fit with the general pattern might be treated separately. We shall return to this point in Subsection 2.4.

2.1 Positive and negative relationships

Look again at the scatterplot given in Figure 2 (Subsection 1.3), which shows the relationship between percentage of men unemployed and percentage of households with no car.

The points on the scatterplot do not lie exactly on a straight line. This means that if we were told the percentage of unemployed men in a town, we would not know the exact percentage of households without a car. However, knowing the percentage of unemployed men does tell us something about the percentage of households without a car. As was noted in Activity 4 (Subsection 1.3), there is a tendency for towns with a low unemployment rate to also have a low percentage of households with no car. Similarly there is a tendency for towns with a high unemployment rate to have a high percentage of households with no car.

This is more clearly seen by looking at the shaded area shown in Figure 3. The shaded area is chosen so that it contains all the points.

By concentrating on the shaded area instead of the individual points, the general pattern between the percentage of men unemployed in a town and the percentage of households without a car becomes clearer. The area slopes upwards from left to right, so towns with a low unemployment rate, like Mole Valley, have a low percentage of households with no car, while towns that have a high unemployment rate, like Rotherham, also have a high percentage of households with no car.

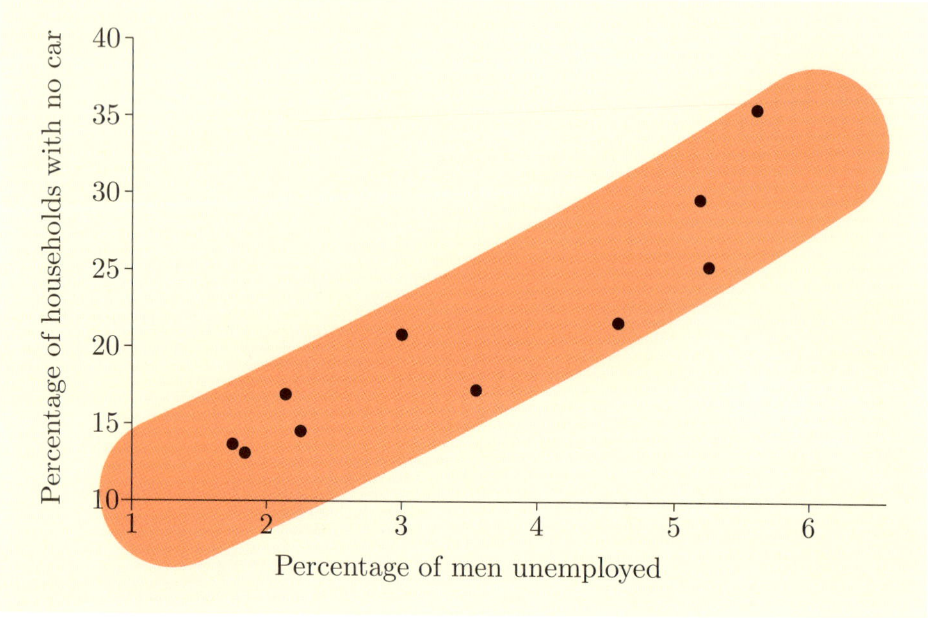

Figure 3 Percentage of males unemployed and percentage of households with no car in ten towns

When the area enclosing the points slopes upwards from left to right, as in Figure 3, then we say that the variables are **positively related**. So, Figure 3 implies that the male unemployment rate and the percentage of households without a car are positively related.

Lots more positive relationships

Figure 4 shows the scatterplot of some data relating to weekly household expenditure for 12 regions and nations in the UK. The data points are again enclosed in a shaded area.

This time the area slopes downwards from left to right. High weekly expenditure is associated with a low percentage of expenditure on food and non-alcoholic drink, and low weekly expenditure is associated with a high percentage of expenditure on food and non-alcoholic drink. When large values of x are usually associated with small values of y, and small values of x are associated with large values of y, as in Figure 4, the variables are said to be **negatively related**.

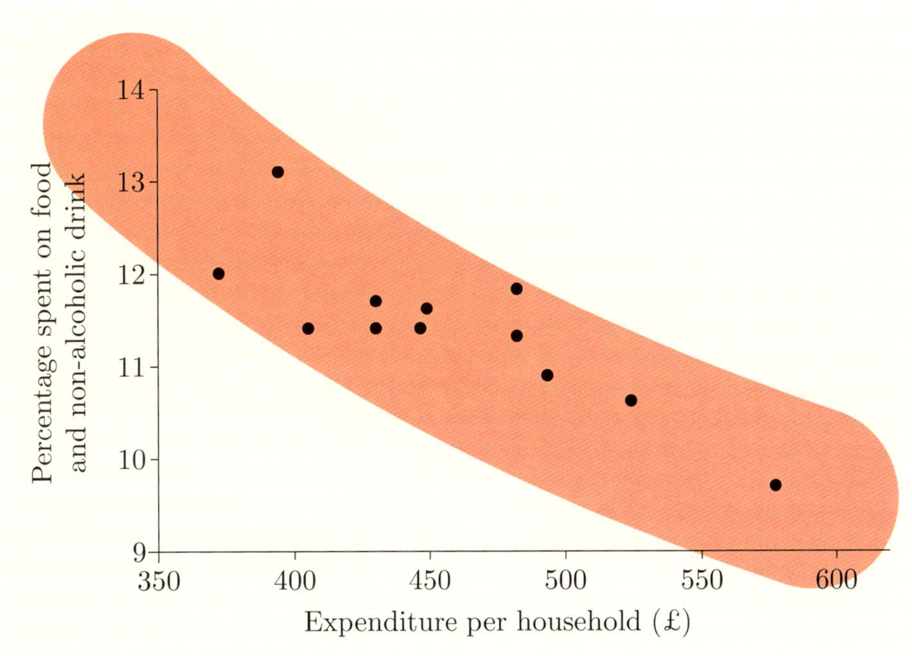

Figure 4 Expenditure per household and percentage spent on food and non-alcoholic drink

(Data source: ONS (2011) *Family Spending: 2010 edition*, Table A33)

Positive and negative relationships

On a scatterplot, variables are said to be **positively related** if low values of x are associated with low values of y, and high values of x are associated with high values of y.

That is, if points tend to slope *upwards* from left to right, then the variables are *positively* related.

Variables are said to be **negatively related** if low values of x are associated with high values of y, and high values of x are associated with low values of y.

That is, if points tend to slope *downwards* from left to right, then the variables are *negatively* related.

Activity 6 *Positive or negative relationship?*

In Figures 5 and 6 below, are the variables positively or negatively related?

(a) A dataset of 50 observations.

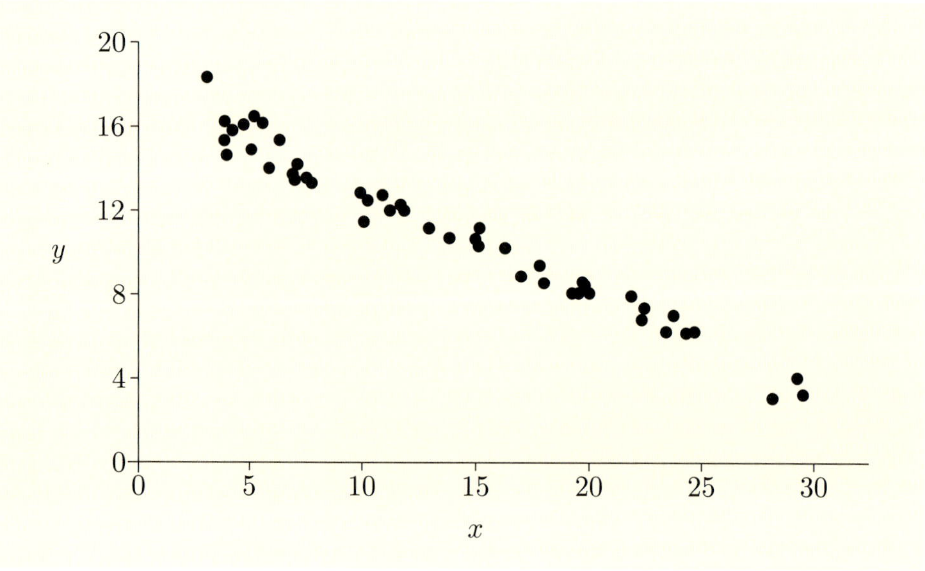

Figure 5 A scatterplot of some data

(b) A scatterplot of data from an experiment in kinesiology. A subject performed a standard exercise task at a gradually increasing level. The x-coordinate measures the amount of oxygen uptake, and the y-coordinate is the expired ventilation, which is related to the rate of exchange of gases in the lungs.

Note that the units for the oxygen uptake and expired ventilation would normally be included in the labelling of the scatterplot axes. However, in this case, the units are not known. The precise choice of units does not make a difference as to whether the relationship is positive or negative.

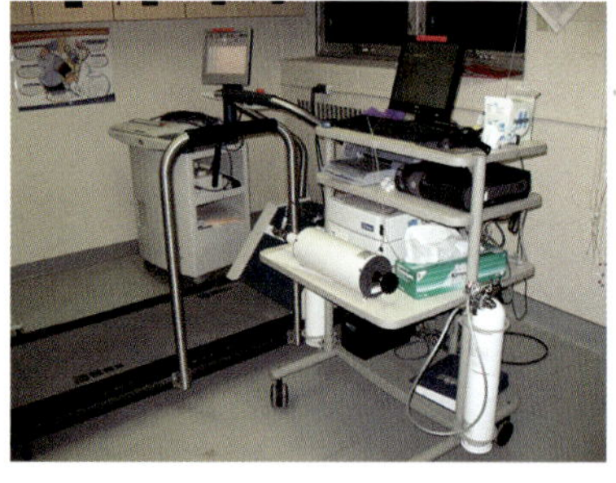

An example of the type of equipment that can be used to measure oxygen uptake

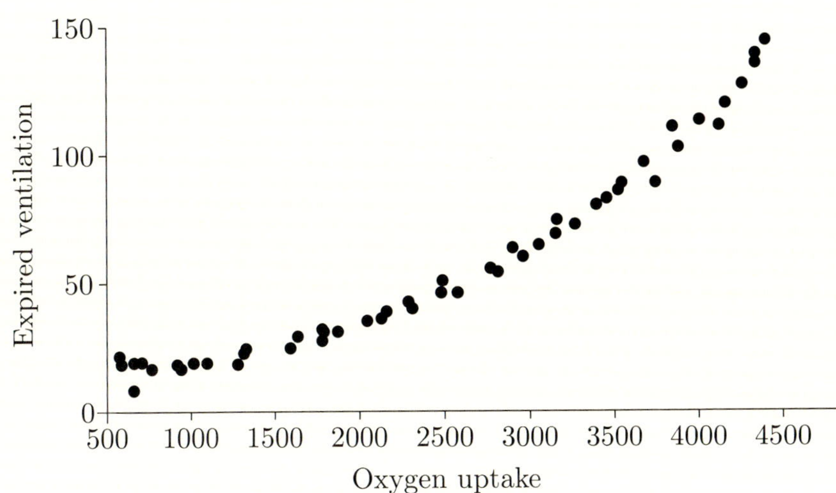

Figure 6 Data from an experiment in kinesiology

(Data source: Bennett, G.W. (1988) 'Determination of anaerobic threshold', *Canadian Journal of Statistics*, vol. 16, no. 3, pp. 307–310)

Not all pairs of variables are positively or negatively related. Figure 7 shows the number of adult road-user casualties in Scotland on weekdays, averaged over 2005 to 2009, for each hour of the day.

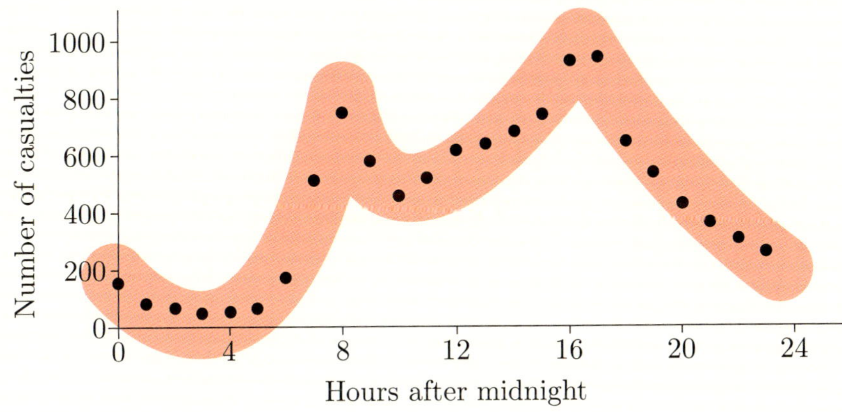

Figure 7 Casualties to adult road users on weekdays

(Data source: The Scottish Government (2010) *Reported Road Casualties – Scotland 2009*, Table 28)

You can see that there is a very pronounced pattern; the number is highest between 4 pm and 6 pm (16:00 and 18:00) when many adults are going home. The number is also quite high between 8 am and 9 am when many adults are travelling to work or doing the school run. On the other hand, the number is very low between 12 am and 7 am.

So there is a definite relationship between time of day and number of casualties, but it is much more complex than one that can be described simply as either positive or negative.

2.2 Linear and non-linear relationships

In the previous subsection, you saw that relationships on scatterplots can be described as positive, negative or neither. This subsection concentrates on another aspect of a relationship that can be investigated by looking at a scatterplot: whether it is linear or non-linear.

Example 4 *Daily electricity usage and cost*

In 2011, a household recorded the cost of their electricity usage on ten different days. The number of kilowatt hours (kWh) they used on each of these days along with the amounts they were charged are shown in Figure 8.

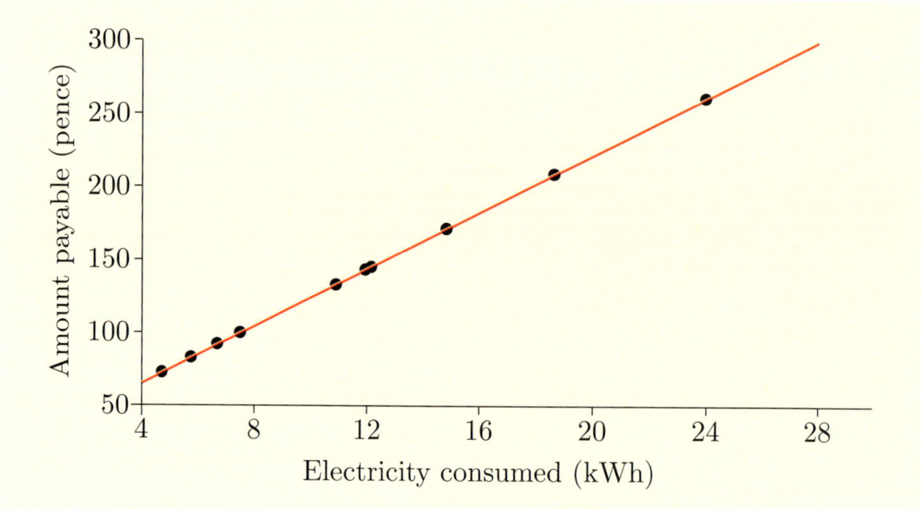

Figure 8 Daily cost of electricity for a household

Looking at the scatterplot, it is clear that there is a very precise relationship between the quantity of electricity consumed and the amount payable on the electricity bill: the data points lie on a straight line.

Many domestic electricity tariffs in the UK are made up of a fixed standing charge per day plus a charge per kilowatt hour used. For this household, the standing charge was 27 pence per day, and the cost per kWh was 9.7p. On day 1, this household's electricity cost $11.96 \times 9.7\text{p} = 116\text{p}$ (rounded to the nearest penny), so the total amount payable for that day was $27\text{p} + 116\text{p} = 143\text{p}$. If a household uses x kWh of electricity in a day, the total payable for that day is $27 + 9.7x$ pence. So if y is the amount payable in pence, then

$$y = 27 + 9.7x.$$

This is the equation of the straight line shown in Figure 8.

The relationship between electricity usage and amount payable in Figure 8 is represented by a straight line. So there is said to be a **linear relationship** between the electricity used and the amount payable.

The points need not lie exactly on a straight line for a relationship to be linear. Consider again the relationship between male unemployment and the percentage of households with no car. The shaded area of the scatterplot in Figure 3 (Subsection 2.1), and hence the underlying relationship, can be summarised by drawing a line through the middle of it as shown in Figure 9. This line happens to be a straight line, so this relationship is also said to be linear. More precisely, it is said to be a **positive linear** relationship, as the line goes up from left to right.

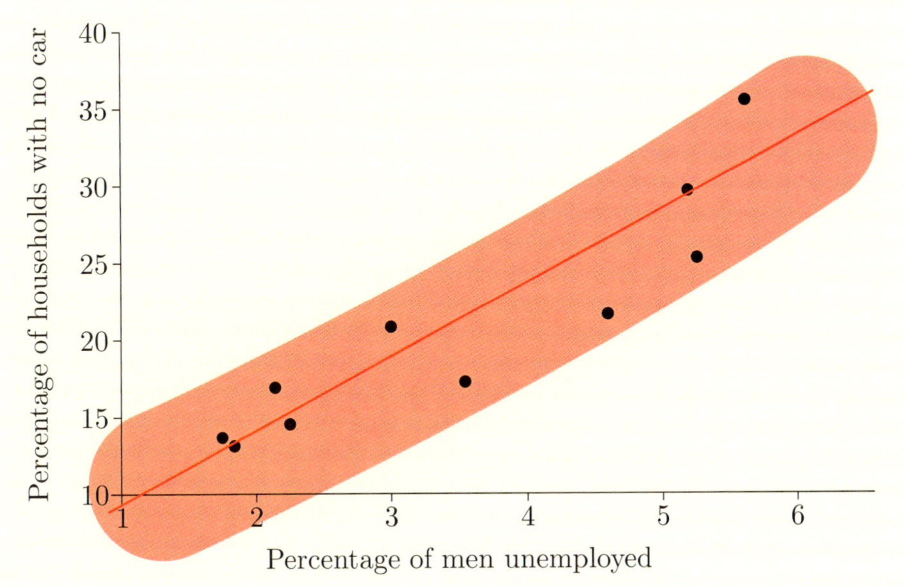

Figure 9 Percentages of males unemployed and households with no car in ten towns, with a line in the middle of the shaded area

It is also possible to draw a straight line going through the shaded area on the scatterplot of household expenditure given in Figure 4 (Subsection 2.1). However, this time the line would go down from left to right. So the relationship between household expenditure and the percentage spent on food is a **negative linear** relationship.

Now consider again the data from the experiment in kinesiology introduced in Activity 6 (Subsection 2.1). In Figure 10 a line has been drawn through the middle of the shaded area covering all the points.

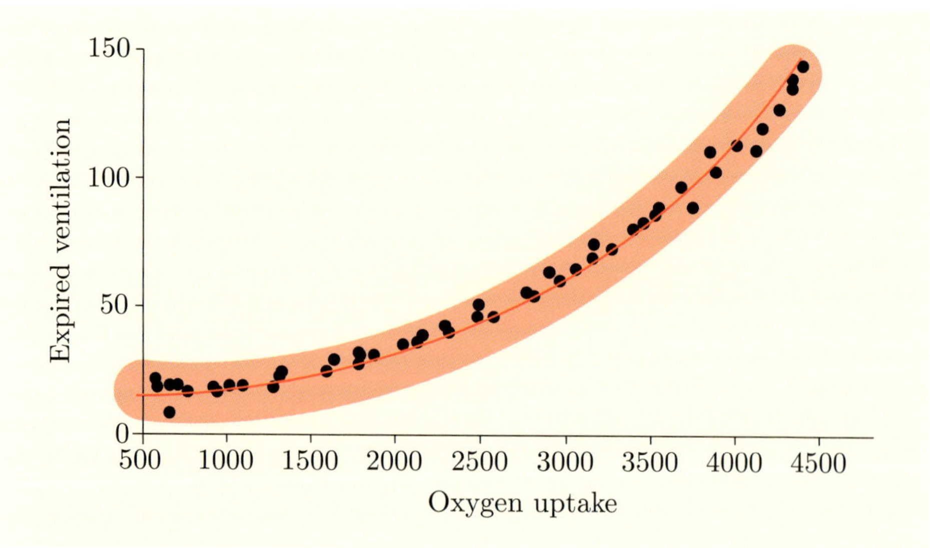

Figure 10 Data from an experiment in kinesiology with a line in the middle of the shaded area

Notice that the line in Figure 10 is curved, not straight. This is because the area containing the points is distinctly curved. So we say that the relationship between oxygen uptake and expired ventilation is **non-linear**.

Linear and non-linear relationships

A relationship is said to be linear if it can be summarised reasonably well by a straight line.

A relationship is said to be non-linear if it can be summarised reasonably well by a curve but not by a straight line.

Activity 7 *Linear or non-linear?*

Figure 11 is a scatterplot of data from an ultrasonic calibration study. Is the relationship between the variables linear or non-linear?

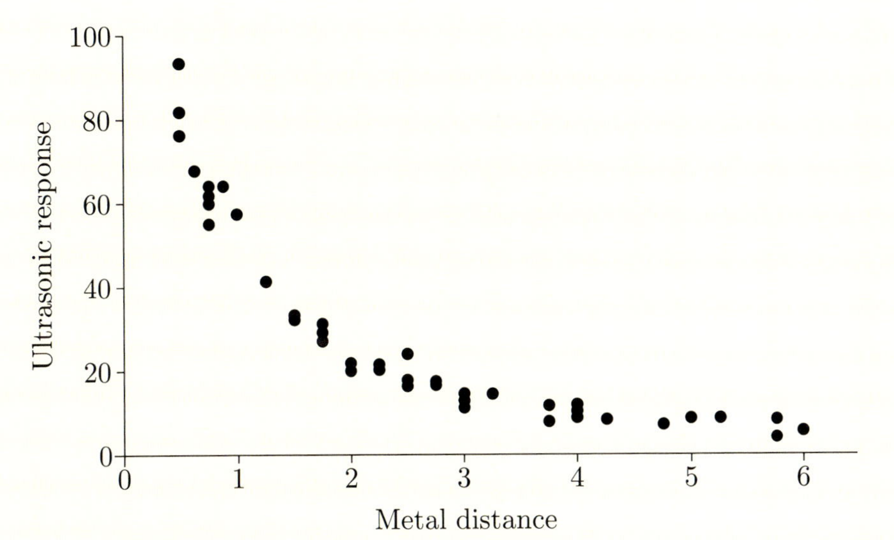

Figure 11 Data from an ultrasonic calibration study

(Source: Castillo, E., Hadi, A.S. and Minguez, R. (2009) 'Diagnostics for non-linear regression', *Journal of Statistical Computation and Simulation*, vol. 79, no. 9, pp. 1109–1128)

2.3 Strong and weak relationships

Sometimes the general pattern formed by points on a scatterplot is very clear: if a line were drawn on the plot summarising the general pattern, then all the points would lie close to the line. In such cases we say that there is a **strong relationship** between the two variables. On the other hand, the general pattern might be difficult to pick out. Then the relationship between the two variables is said to be a **weak relationship**.

Example 5 *A strong relationship*

Look again at the data from the kinesiology experiment introduced in Activity 6 (Subsection 2.1). This scatterplot is reproduced in Figure 12 for convenience.

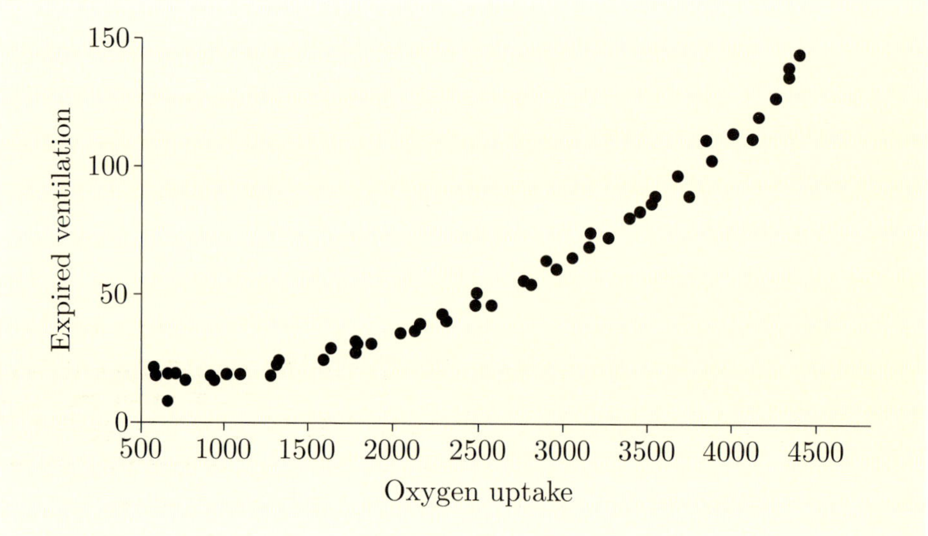

Figure 12 Data from an experiment in kinesiology

Notice that the points form a clear general pattern, curving upwards as you move from left to right. All the points lie close to this general pattern. Thus there is a strong relationship between oxygen uptake and expired ventilation.

Example 6 *A weak relationship*

Data were collected on the water consumption of customers from 22 water companies in the UK in 2008/09. The data are plotted in Figure 13.

Notice that in this scatterplot the overall pattern is not very clear. The points go generally up as you move from left to right. However, there is lots of scatter around whatever trend there is. So there is a weak relationship between the average consumption in metered households and the average consumption in unmetered households served by the same company.

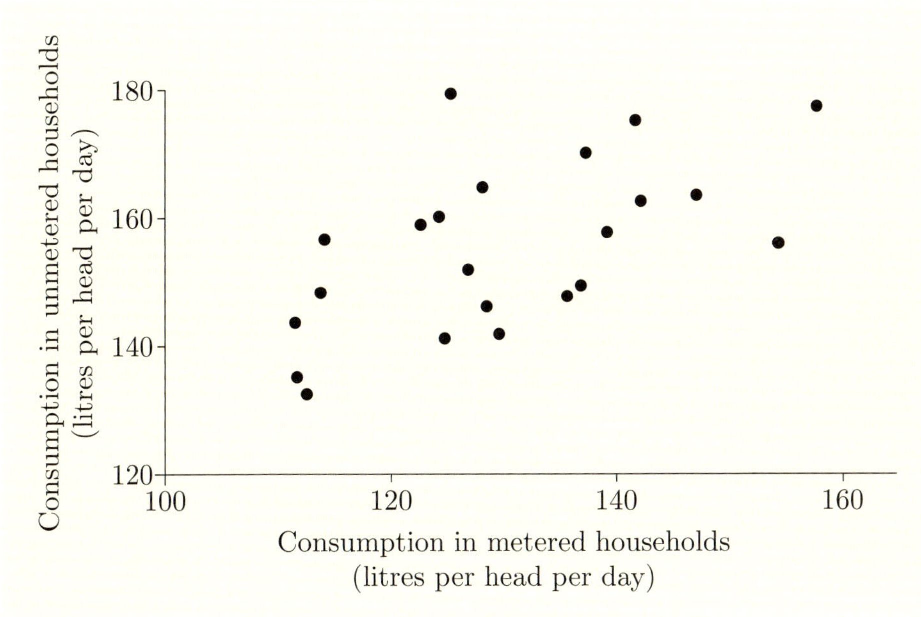

Figure 13 Average water consumption in metered and unmetered households

(Data source: ONS (2011) 'Regional Trends Online Tables', Table 5.4)

Strong and weak relationships

A relationship is said to be **strong** when all the points on a scatterplot lie close to a line.

A relationship is said to be **weak** when the points only loosely follow a line.

Deciding when a relationship is clear enough to be a 'strong' relationship is a subjective judgement. Sometimes the best that can be done is just to say whether a relationship on one scatterplot looks stronger than the relationship on another plot.

Activity 8 *Comparing strength of relationships*

Order the scatterplots in Figures 14, 15 and 16 according to the strength of the relationship between the variables, from strong to weak.

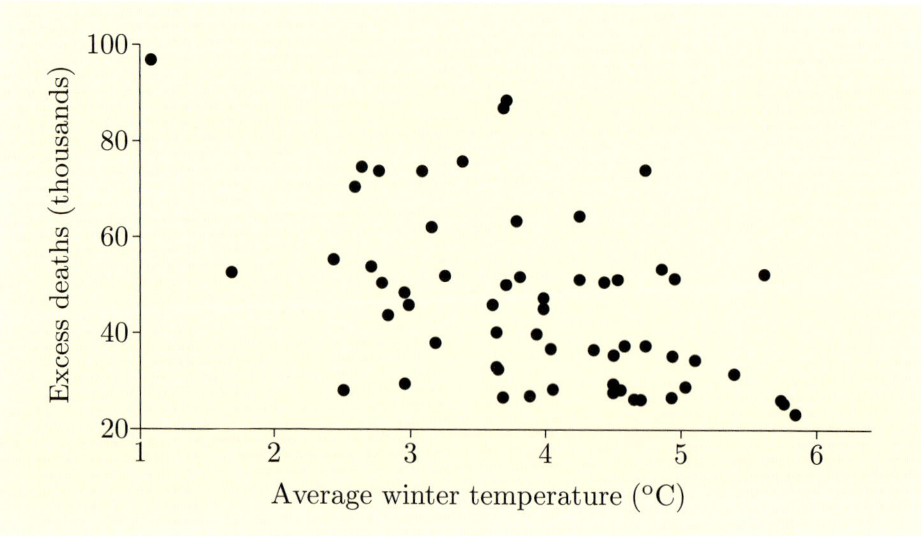

Figure 14 Estimated number of 'excess' deaths in winter (over and above the average for the rest of the year), plotted against average winter temperature, for Great Britain for each year from 1952 to 2010.

(Data source: HMSO (2011) *Social Trends 41 – Health*, data for Figure 6)

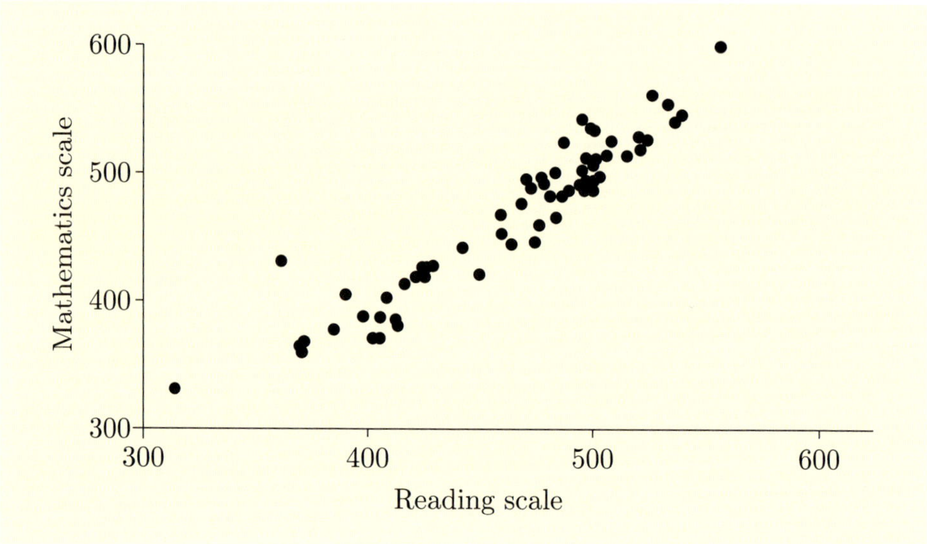

Figure 15 Average performance on a mathematics scale plotted against average performance on a reading scale, for 15-year-old students from different countries in 2009.

(Data source: OECD (2010) *PISA 2009 Results: What Students Know and Can Do – Student Performance in Reading, Mathematics and Science (Volume I)*)

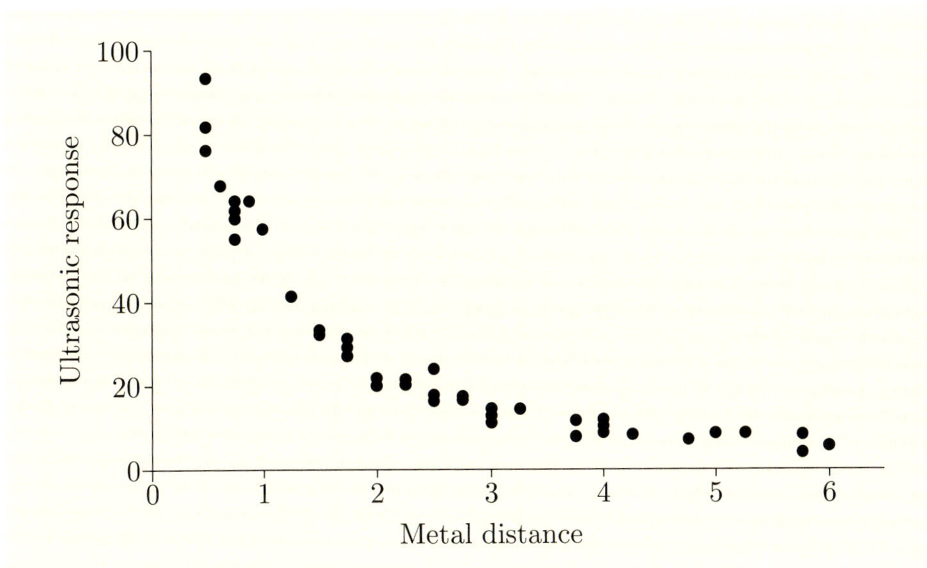

Figure 16 Data from the ultrasonic calibration study introduced in Activity 7 (Subsection 2.2)

Activity 9 *Describing a relationship*

Look at the data displayed in Figure 17. This scatterplot displays further information about the ten towns listed in Table 3 (Subsection 1.2) and considered in Figure 9 (Subsection 2.2). The x-axis corresponds to the percentage of employed residents working in the manufacturing industry, and the y-axis corresponds to the percentage of households living in owner-occupied houses.

What relationship between the variables can you observe in this scatterplot?

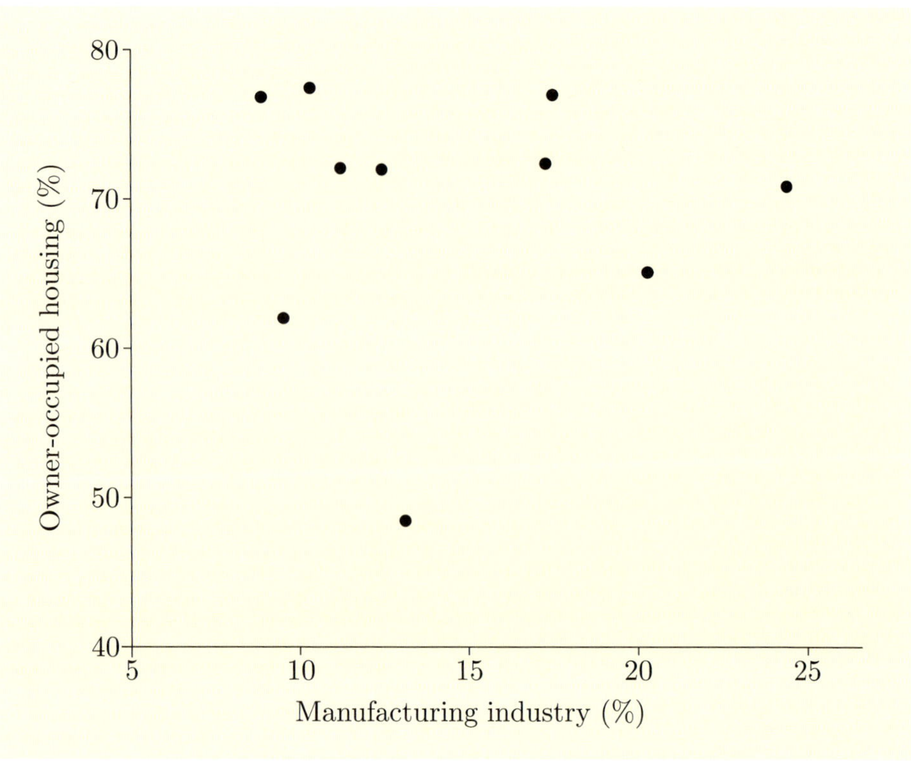

Figure 17 Percentage of employed residents working in manufacturing and percentage of households in owner-occupied houses

(Data source: HMSO (2003) *Census 2001: Key Statistics for Local Authorities in England and Wales*, Tables KS11a and KS18)

The scatterplot given in Activity 9 is an example of no relationship between two variables. If you were told, for example, that 14.4% of the workforce of Milton Keynes were employed in manufacturing industries, this would not help you at all in estimating the proportion of households who own their own homes. (Actually, it was 65.2% in 2001.)

There is said to be **no relationship** between two variables when knowledge of the value of the explanatory variable does not provide information about the value of the response variable.

2.4 Unusual points

In this subsection, there are some final comments about interpreting scatterplots and they relate to unusual points: sometimes one or two data points do not appear to follow the same pattern as the rest of the points.

For example, look at Figure 18, which shows a scatterplot of the percentage of the population aged under 16 in different regions of the UK in 2010 and the population densities (in people per km^2).

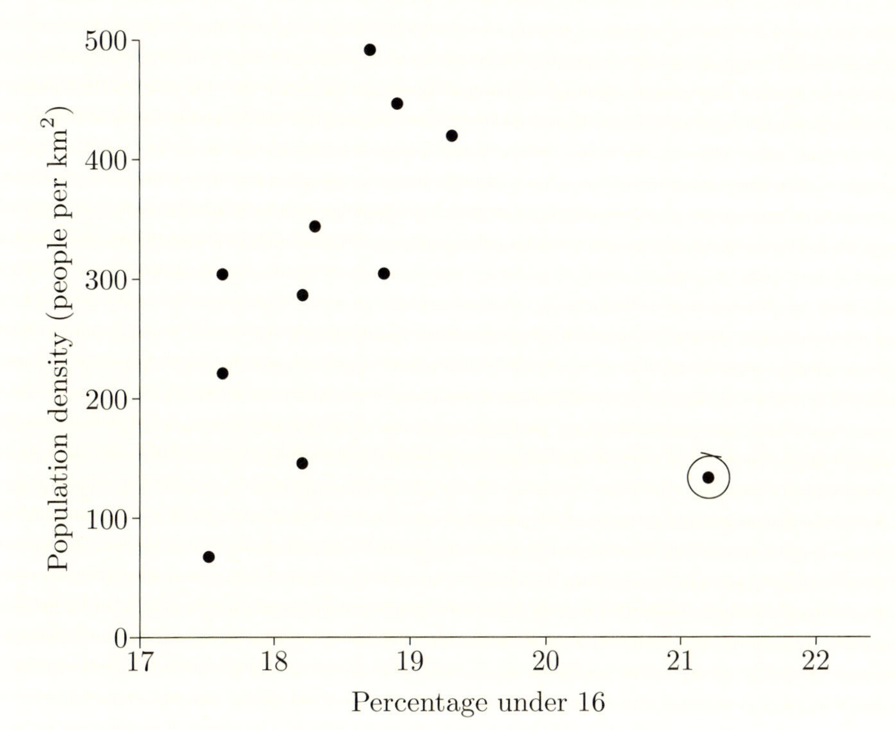

Figure 18 Percentage of the population aged under 16 and the population density

(Data source: ONS (2011) *Region and Country Profiles, Key Statistics – October 2011*)

You can see that the ringed point, which represents Northern Ireland, does not follow the general pattern of the other ten regions. This point is called an *outlier*, because it is inconsistent with the main body of the data. This extends the definition of *outliers* given in Subsection 4.2 of Unit 1, where we considered only one variable at a time. The particular reasons for the investigation would determine whether or not Northern Ireland should be included when summarising the relationship between the two variables.

The x-value for Northern Ireland is unusual as it is much larger than those of all the other points. More generally, a point can be inconsistent with the main body of data even though neither its x-value nor its y-value is unusual – the *combination* of its x- and y-values can still place it a long way from other points and make it an outlier.

Finally, it should be noted that sometimes a point is only an outlier because a mistake has been made in the observation and/or recording of a data point. For example, if in the data used for Figure 18, the percentage of the population aged under 16 should really have been 18.2% instead of 21.2%, the point representing Northern Ireland would no longer appear to be an outlier. So looking for outliers on a scatterplot can help in data cleaning by highlighting parts of the data that are worth checking again for errors.

Data cleaning was introduced in Subsection 3.1 of Unit 1.

Activity 10 *Spot the outlier*

For each of the scatterplots below, how many outliers can you identify?

(a) The scatterplot of the average performance of 15-year-olds in different countries in 2009, introduced in Activity 8 (Subsection 2.3).

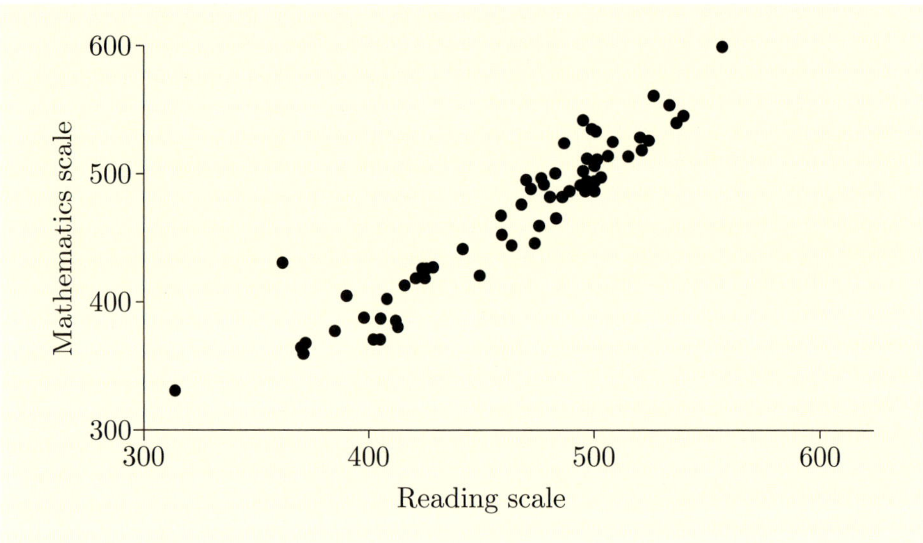

Figure 19 Student performance in reading and mathematics

(b) The scatterplot of some data relating to weekly household expenditure for 12 regions and nations in the UK, introduced in Figure 4 (Subsection 2.1).

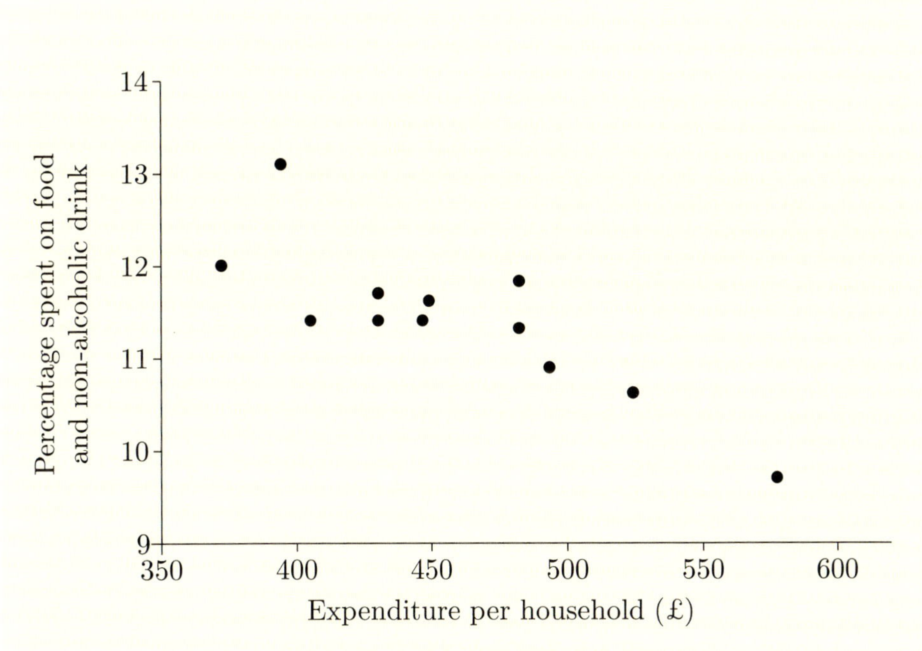

Figure 20 Percentage spent on food by households

In this section you have been learning to interpret scatterplots. Here is a checklist of things to consider.

Checklist for interpreting scatterplots

- Is the relationship positive, negative or neither?
- Is the relationship linear or non-linear?
- Is the relationship strong or weak?
- Are there any outliers?

You have now covered the material related to Screencast 1 for Unit 5 (see the M140 website).

Exercises on Section 2

Exercise 3 *Average wages of men and women*

Figure 21 is a scatterplot of men's and women's average hourly wages in different sectors of the UK economy. Interpret this scatterplot.

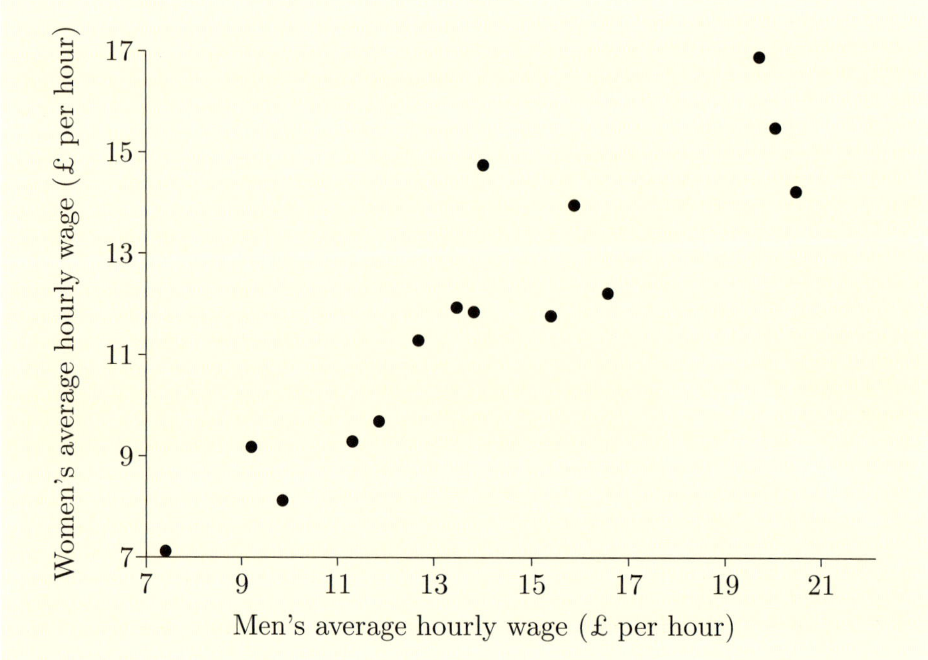

Figure 21 Average hourly wage for men and women in 15 sectors of the UK economy

(Data source: ONS (2012) *Average Weekly Earnings dataset: September 2012*)

Exercise 4 *Investigating house prices over time*

In Figure 22, a scatterplot of the average house price in the UK over the period 1991 to 2008 is shown. Using this scatterplot, comment on the pattern of house prices over this period.

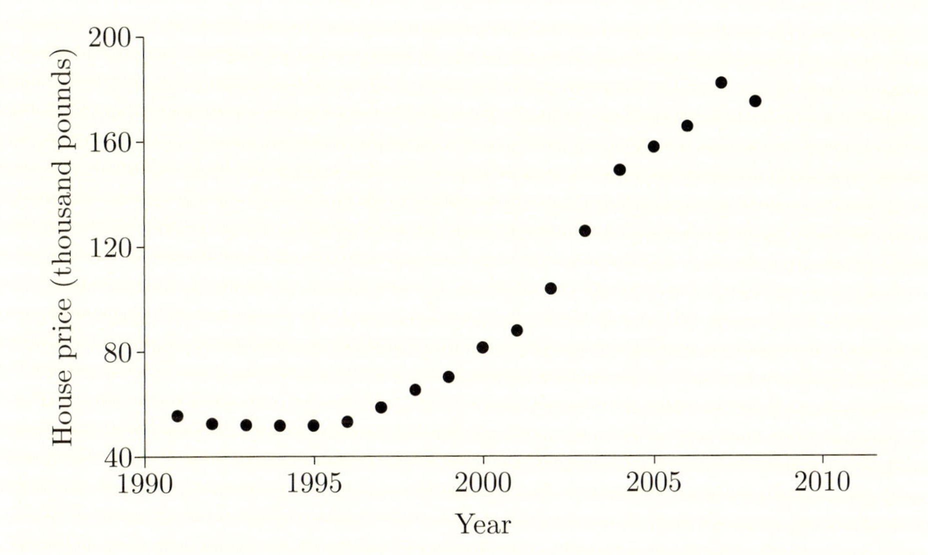

Figure 22 Average house price in the UK between 1991 and 2008

(Data source: Nationwide building society (2013) 'UK house prices since 1952')

3 Scatterplots and lines

In Section 1, you were introduced to the idea of a relationship between two variables and learned how to present a relationship as a scatterplot. In Section 2, you saw how relationships may be positive, negative or neither, and that they can be linear or non-linear. In this section and Section 4, you will learn how to describe a relationship by adding a line and by calculating an equation. The line has many uses. For example, it will allow us to say something about the rate at which the response variable changes as the explanatory variable changes, and also to make informed predictions about the response when the value of the explanatory variable is known.

3.1 Drawing lines

You saw in Subsection 2.1 that many scatterplots could be approximately summarised by outlining an area on the scatterplot in which all, or nearly all, of the points lie. Often this area is long and narrow; for many datasets it is roughly straight, though it may be curved, as in Figure 6. However, sketching an area gives only a vague summary, and it would be useful to have a more precise measure.

In Example 4 (Subsection 2.2), we saw that the data points for daily electricity costs all lay exactly on a straight line. The line $y = 27 + 9.7x$ provides a precise summary of the data, and this line represents the relationship. In an **exact relationship** between two variables all the points lie exactly on a line which is either straight or follows a simple curve. Can we also represent an inexact relationship using a line? Well, often we can, just as we can represent the location of a batch of data by the median or the mean. Usually, hardly any of the data points are exactly equal to the mean, and, in the same way, a line used to represent a relationship will pass through very few, if any, of the data points. The purpose of the line is to represent the pattern that we can see in the points.

In statistics, the process of finding a line that best represents a relationship is known as **regression**.

Example 7 *Summarising unemployment and car ownership*

Data on percentages of males unemployed and households with no car were introduced in Subsection 1.2. In Subsection 2.2, you saw that the relationship between these quantities is approximately linear because the data can be summarised reasonably well by a straight line. One such line is shown in Figure 23. (Ways of choosing such a line will be considered later in this subsection and in Section 4.)

The line highlights the fact that towns with high male unemployment tend to have a relatively high percentage of households with no car, while those with low male unemployment tend to have a relatively low percentage of households with no car. The equation of the line shown in the figure is $y = 5.8 + 4.2x$.

Figure 23 Percentage of males unemployed and percentage of households with no car, with straight line

Example 8 *Summarising oxygen uptake*

Recall from Subsections 2.1 and 2.2 that the scatterplot of oxygen uptake suggests a positive non-linear relationship between oxygen uptake and expired ventilation. This suggests that the data can be summarised by a curve that goes up as you move from left to right. One such curve is shown in Figure 24.

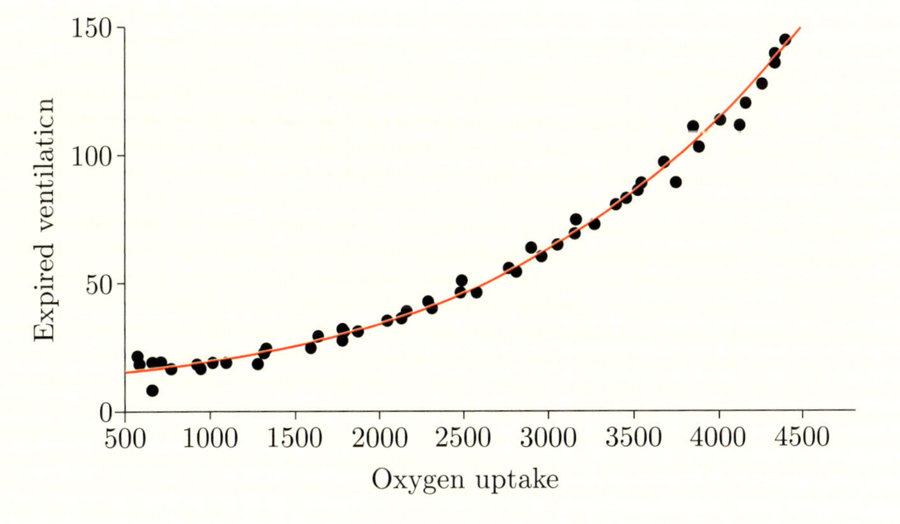

Figure 24 Data from an experiment in kinesiology, with curved line

Notice that the points are generally close to the curve. This is to be expected, as the relationship between oxygen uptake and expired ventilation is a strong one (as noted in Example 5, Subsection 2.3).

The lines in Examples 7 and 8 were both drawn by looking at the scatterplot and choosing a line that appears to provide a sensible model for the pattern made by the points.

Activity 11 *Summarising data with a line*

Suppose that the fictional data depicted in Figure 25 come from an experiment. In this experiment, an industrial process was run seven times, each time at a different temperature, and the yield recorded.

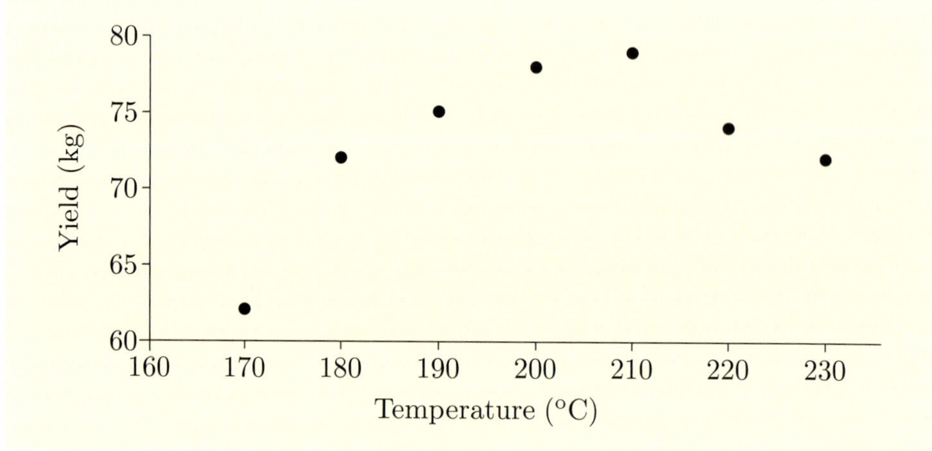

Figure 25 Yield from an industrial process

(a) Which variable is being treated as the response variable, and which as the explanatory variable? Does this choice seem reasonable?

(b) Briefly describe the relationship between temperature and yield shown in the scatterplot.

(c) Using your judgement, draw a line on Figure 25 which you feel provides a good summary of the data.

Your attempt at a line was probably slightly different from the one given in the solution to Activity 11(c). There is no single right answer to this question. However, your curve was probably of the same general inverted U-shape. The data certainly suggest that the yield is greatest when the temperature is round about 200–210 °C, but less if the temperature is either cooler or hotter. The line could have been drawn as a wiggly curve that went through all seven points on the scatterplot as in Figure 26.

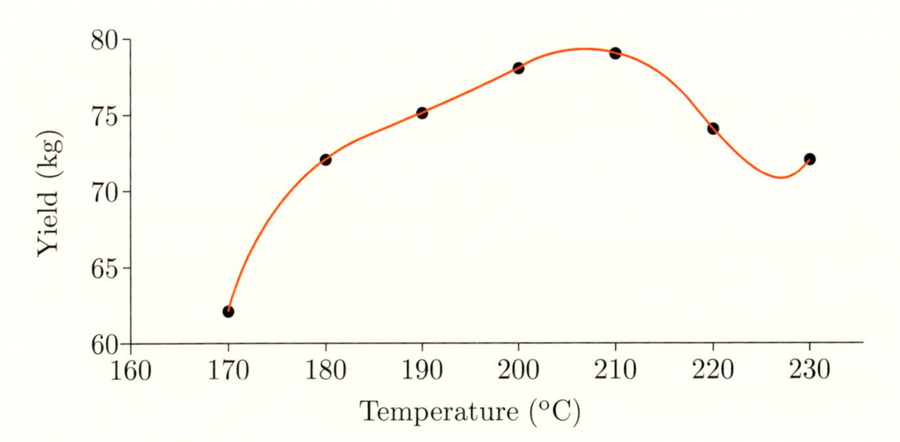

Figure 26 Yield from an industrial process, with a summary line going through all data points

However, it is important to bear in mind that the data were the result of one experiment. If the experiment had been repeated at 170 °C, the yield might have been 61 or 64, say. The yield is subject to some experimental uncertainty, so a curve passing exactly through the points, as shown, is not the best way of summarising the data. A simpler curve is more useful.

To describe a scatterplot, we shall look for a simple curve that summarises the relationship. A straight line is the simplest sort of curve there is, so that will be used when it seems appropriate. For many of the scatterplots that we have looked at so far in this unit, a straight line does provide a reasonable summary of the pattern in the data.

Describing a scatterplot

When summarising data on a scatterplot, the simplest adequate curve should be chosen. In many cases this amounts to choosing an appropriate straight line. This line is called the 'fitted line' or 'fit line'.

The process of choosing a straight line to draw is often called **fitting a line** to the data. There are many different ways to do this. One method, which you have already met, is simply to draw in the line that appears to give a good representation of the pattern in the data. This method, known as **fitting by eye**, can be perfectly adequate, particularly if the relationship is strong and all points are close to the line. But choosing a line that looks 'about right' is less easy when the relationship is weak and the points are widely scattered.

Activity 12 *Comparing lines fitted by eye*

Figure 27 shows four attempts at fitting a line by eye to the same set of data. In each case, say whether you think the line is a good choice or a poor choice. If you think it is poor, suggest how it might be improved.

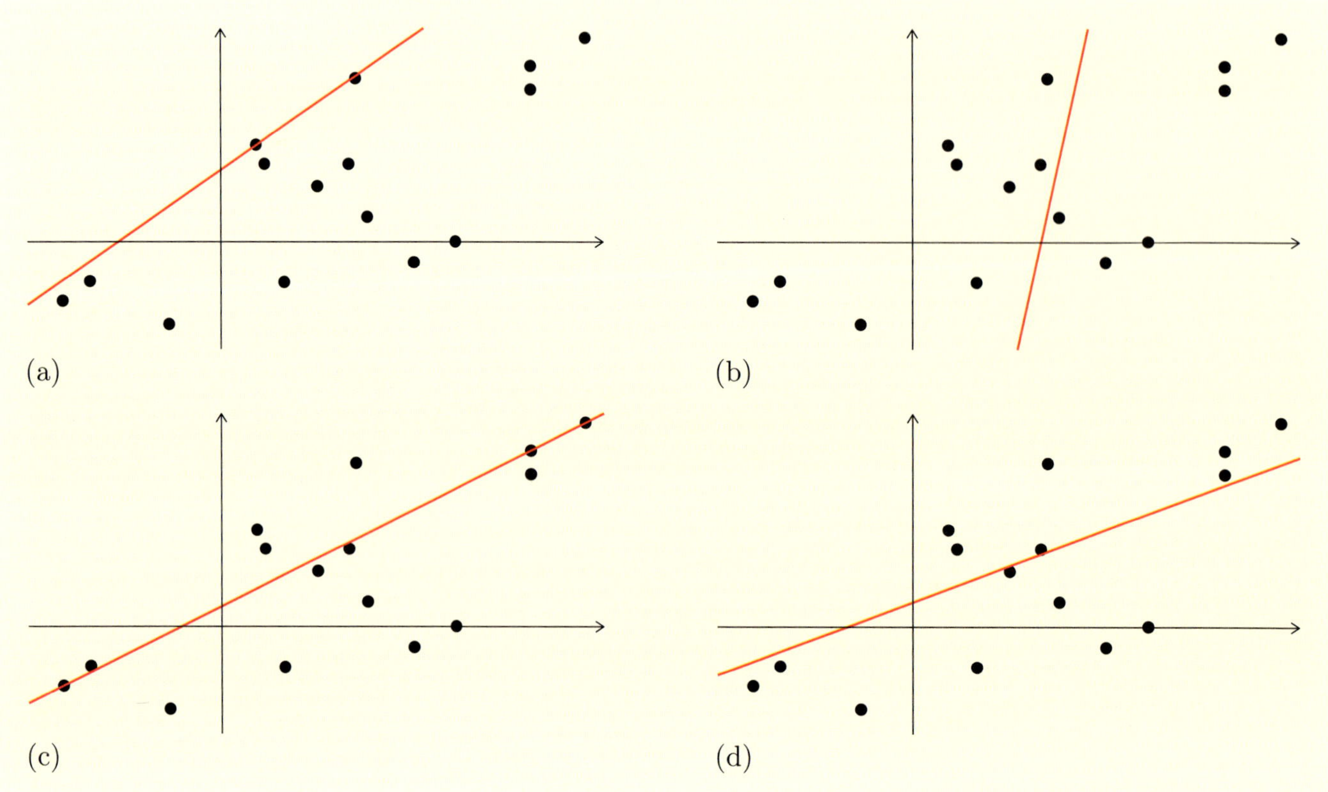

Figure 27

You saw in the solution to Activity 12 that sometimes it is easy to say that a straight line is not a good fit. In Figure 27(a), the line was above all but two of the data points, and it would obviously be better if it had been lower and had passed through the middle of the cluster of points. Neither is the line in Figure 27(b) a good fit. Again, it does not follow the general pattern of the points, and it would be better if it were rotated clockwise to bring it near to the position of the line in either Figure 27(c) or (d). To choose a line that fits the general pattern of the data, it is sometimes helpful to use a transparent ruler and move it around until it appears to be in a good position.

Fitting lines is good for you: the logo of a German purveyor of vitamin and mineral supplements!

However, it was hard to decide whether Figure 27(c) or (d) was a better choice. To get any further with the problem of deciding which straight line to draw, and whether a straight line does provide an adequate summary of the data, we need a more definite idea of what we mean by a *good* summary. We can think of this as whether the line provides a good fit to the data. This idea will be developed in the next subsections.

3.2 Residuals

The basic idea for residuals uses an equation that will reappear later in this module: the **DFR equation**.

The DFR equation

This equation splits an observed 'Data' value into two parts: the 'Fit' and the 'Residual'. These are linked in the following way.

$$\text{Data} = \text{Fit} + \text{Residual}.$$

The equation can be rearranged as

$$\text{Residual} = \text{Data} - \text{Fit}.$$

In other words, a residual is defined as the difference between a data value and a fit value.

Now suppose that when there are linked data, the 'Data' is taken to be the response variable. That is, for every point on the scatterplot, the 'Data' is the position of that point up the y-axis. And suppose that the 'Fit' is taken to be the position of a fitted line along the y-axis. That is, for every point on the scatterplot, the 'Fit' is the vertical position of the line, for the value of the explanatory variable. Then the 'Residual' is a measure of how far away each data point is from the fitted line. This is illustrated in Figure 28.

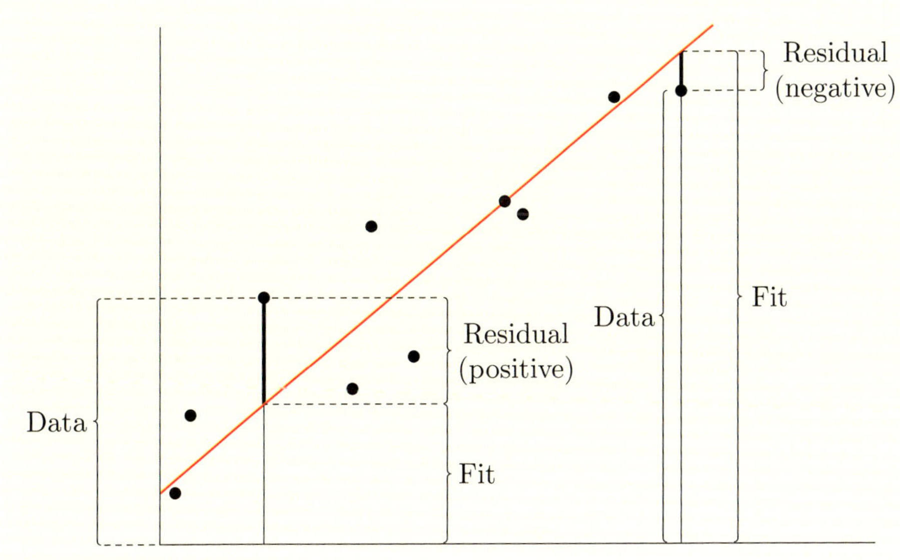

Figure 28 A scatterplot showing two residuals

Note that a 'Residual' is the *vertical* distance between a data point and the line. The 'Residual' is positive if the line is below the data point, and the 'Residual' is negative if the line is above the data point. If the 'Residual' is zero, the data point lies exactly on the fitted line. A 'Residual' close to zero indicates that the point is close to the fitted line, and a 'Residual' further away from zero indicates that the point is a long way from the line.

The reason for focusing on vertical distances is clearer when values of the explanatory variable are fixed by the experimenter. In the introduction to this unit, an experiment from Subsection 2.1 of Unit 1 was mentioned, in which fertiliser was applied and the subsequent yields of grain were recorded. The levels of fertiliser used in the experiment corresponded to at the levels of 0, 25, 50, 75, 100 and 125 kg/ha. In Figure 29, a vertical dashed line is plotted through each of these levels of fertiliser. Even while the grain is growing, we know a scatterplot will put a value for yield on each of these vertical lines.

In Figure 30, the data points have been added, together with the fitted line. The yields we should have expected while the grain was growing are the points where the vertical lines cross the fitted line – these points are the 'Fit' values. The short, thicker lines from the fitted line to the data points are the residuals.

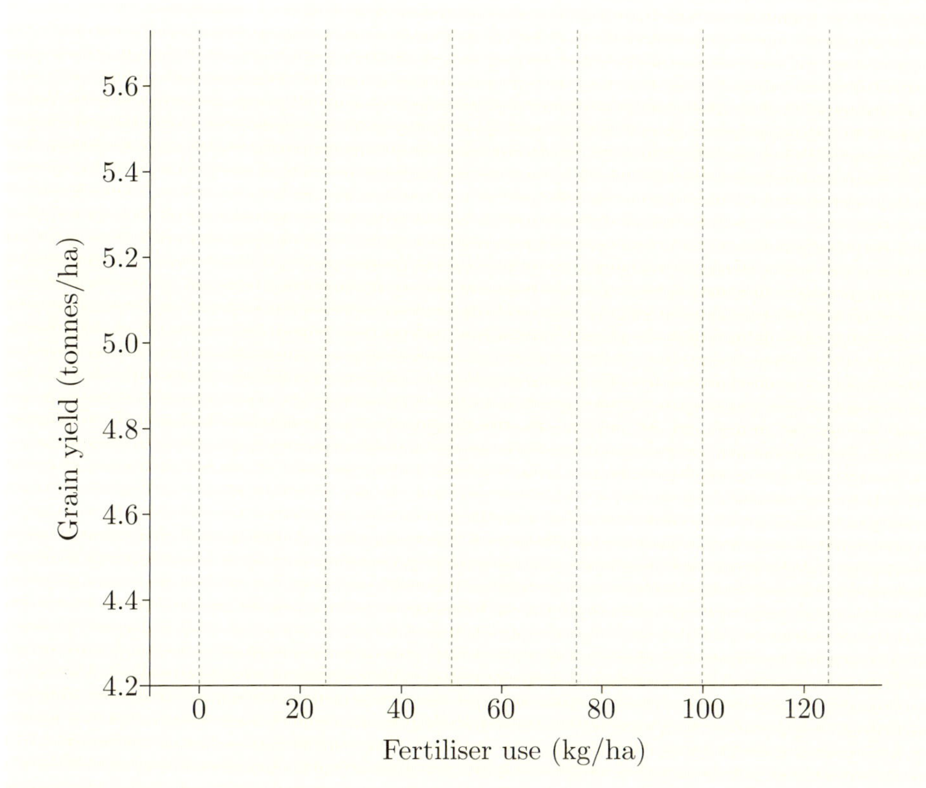

Figure 29 The levels of fertiliser that were applied

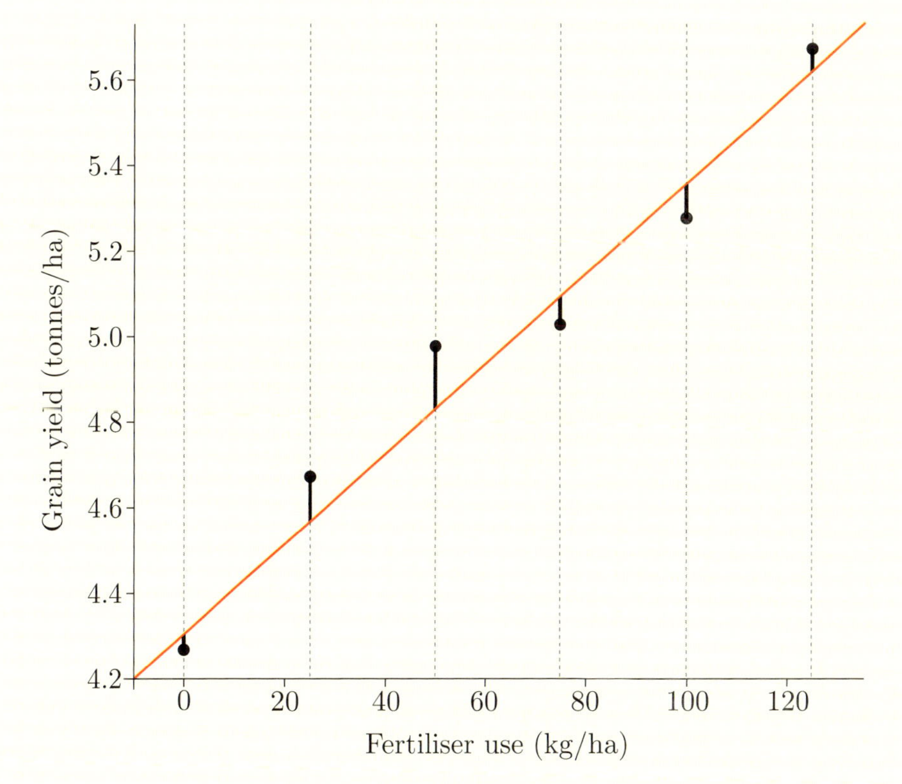

Figure 30 Residuals for the grain data

Example 9 *Calculating residuals*

Figure 31 is a scatterplot of linked data on the average expenditure per week and the average percentage spent on food and non-alcoholic drink, for 12 households. A line fitted by eye is also shown.

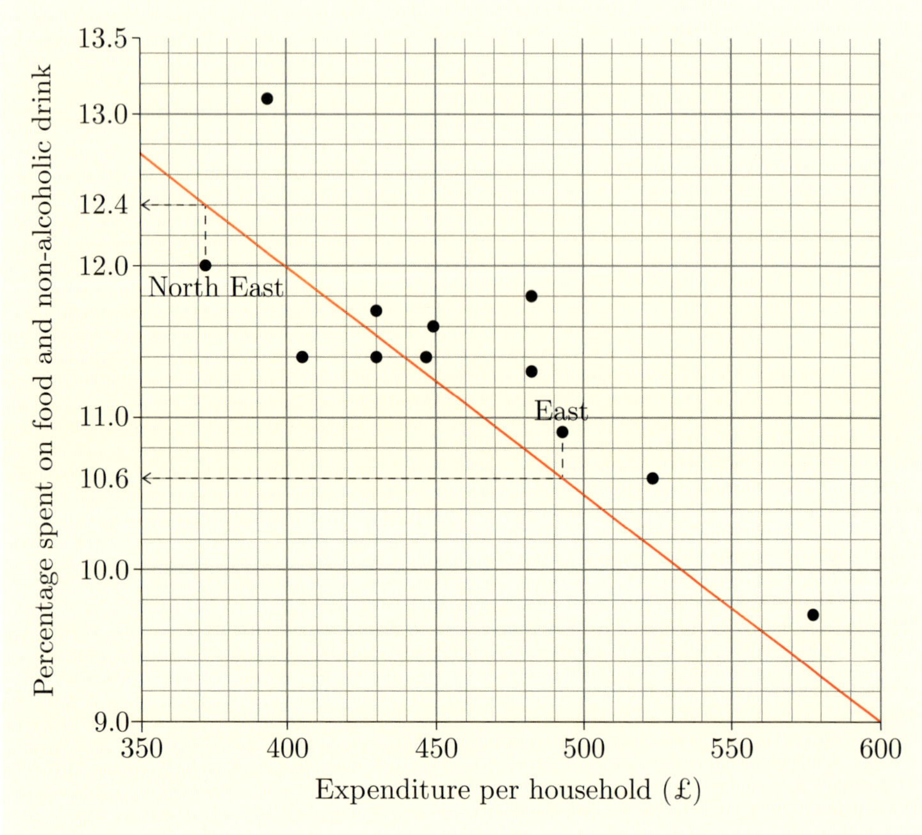

Figure 31 Average weekly household expenditure with a line fitted by eye

Consider the point in the figure that is labelled 'North East'. This point has coordinates $(372.7, 12.0)$. That is the average expenditure per household is £372.70, and the average percentage spent on food and non-alcoholic drink is 12.0%. As the average expenditure per household is the explanatory variable, we are interested in the average percentage spent on food and non-alcoholic drink, y, when $x = 372.7$. In Figure 31, we can see that the point on the line which corresponds to $x = 372.7$ is $(372.7, 12.4)$. Thus 12.4 is the 'Fit' value. Using the DFR equation and applying it to the North East, we find

$$\begin{aligned}\text{Residual} &= \text{Data} - \text{Fit}\\ &= 12.0 - 12.4\\ &= -0.4.\end{aligned}$$

So the 'Residual' for the point corresponding to the North East is -0.4.

Similarly the point labelled 'East' has coordinates $(493.4, 10.9)$, and its 'Fit' value from the scatterplot is 10.6. So for the East,

$$\begin{aligned}\text{Residual} &= \text{Data} - \text{Fit}\\ &= 10.9 - 10.6\\ &= +0.3.\end{aligned}$$

The point representing the North East lies below the line, so its residual is negative. The point representing the East lies above the line, so its residual is positive.

Activity 13 *Reading fit values and residuals from a scatterplot*

A table of all the data plotted in Figure 31 is given in Table 4. Find the fit values and residuals for the rest of the points on the scatterplot in Figure 31. (The fit values and residuals for the North East and the East have already been entered in the table. These were obtained in Example 9.)

Table 4 Weekly household expenditure and percentage spent on food and non-alcoholic drink

Region	x	y	Fit	Residual
England				
North East	372.7	12.0	12.4	−0.4
North West	430.5	11.4		
Yorkshire and the Humber	405.5	11.4		
East Midlands	449.4	11.6		
West Midlands	430.1	11.7		
East	493.4	10.9	10.6	+0.3
London	577.8	9.7		
South East	523.8	10.6		
South West	482.6	11.3		
Wales	394.0	13.1		
Scotland	447.2	11.4		
Northern Ireland	482.8	11.8		

You will have found it quite difficult to measure the fit values accurately using the scatterplot (Figure 31) printed here, as the scale is very small. There is a better method, which involves using the equation of the (fitted) line.

You may also see the equation of a straight line written down in other forms, for example as $y = mx + c$ or $y = ax + b$. All these forms are essentially the same, with just changes to the letters representing the slope and the intercept.

Calculating fit values

The equation of a straight line has the form

$$y = a + bx,$$

where b is the slope or gradient of the line, and a is its intercept (the value of y when $x = 0$). So, for a point where the value of the explanatory variable is x,

$$\text{Fit} = a + bx.$$

Example 10 *Calculating a fit value using the equation of the line*

It turns out that the line drawn on Figure 31 is given by the equation $y = 18.0 - 0.015x$.

For the point representing the North East, $x = 372.7$. The fit value is therefore

$$\begin{aligned} 18.0 - 0.015x &= 18.0 - 0.015 \times 372.7 \\ &= 18.0 - 5.5905 \\ &= 12.4095, \end{aligned}$$

which is 12.4 when rounded to one decimal place. This is the same as the value that we obtained in Example 9 by reading directly from the graph. In general the two values may not turn out to be exactly the same, because of inaccuracies in reading the graph, and rounding of the calculated value.

Activity 14 *Calculating fitted values using the equation of the line*

Calculate the fit value for the following situations.

(a) When the equation of the line is $y = 2 + 4x$ and $x = 12$.

(b) When the equation of the line is $y = -4.6 + 0.3x$ and $x = 3$.

(c) When the equation of the line is $y = -0.5x$ and $x = -2.5$.

(d) When the equation of the line is $y = -3.16 - 4.2x$ and $x = -2.7$.

As has already been noted, using the equation of the fitted line allows fit values to be obtained more accurately than by reading them off from a scatterplot. Also, residual values can then be calculated from the fit values and data values using the DFR equation. This determines their values more accurately than by measuring them directly from the scatterplot.

Activity 15 *Calculating residuals using the equation of the line*

In Example 7 (Subsection 3.1) a straight line was fitted by eye to the data on male unemployment and percentage of households without a car. The equation of this fitted line is $y = 5.8 + 4.2x$, where x is the percentage of males unemployed in a town and y is the percentage of households with no car.

Using the equation of the line, find the residuals for all of the points. For convenience, the data are repeated below.

Table 5 Male unemployment and car ownership for ten towns in England

Town	x	y	Fit	Residual
Alnwick	4.59	21.6		
Vale Royal	3.55	17.2		
Rotherham	5.19	29.7		
Rutland	1.75	13.6		
Dudley	5.27	25.3		
Norwich	5.61	35.5		
Bracknell Forest	2.25	14.5		
Rother	3.00	20.8		
Mole Valley	1.84	13.1		
West Dorset	2.14	16.9		

In this subsection you have learned how to obtain residuals from a line on a scatterplot. In the next subsection you will see how the residuals can be used to help decide whether a line provides a good fit to data points.

You have now covered the material related to Screencast 2 for Unit 5 (see the M140 website).

3.3 Looking for patterns in residuals

We draw a summary line on a scatterplot to try to capture the pattern in the data. If the line is a good fit it should explain all the pattern in the data, and remaining variation around the line should be just random variation. This implies that there should be no pattern in the residuals. If the fit is not a good one, then there may well be some pattern in the residuals.

Example 11 *Looking at a set of residuals: 1*

In Activities 13 and 15, in the previous subsection, you obtained residuals when the line $y = 5.8 + 4.2x$ was fitted to the data on male unemployment and percentage of households without a car. The residuals, ordered by the male unemployment rate, are given again in Table 6.

Table 6 Residuals from a line fitted to the male unemployment and car ownership data

Town	x	Residual
Rutland	1.75	+0.4
Mole Valley	1.84	−0.4
West Dorset	2.14	+2.1
Bracknell Forest	2.25	−0.8
Rother	3.00	+2.4
Vale Royal	3.55	−3.5
Alnwick	4.59	−3.5
Rotherham	5.19	+2.1
Dudley	5.27	−2.6
Norwich	5.61	+6.1

Notice that the residuals appear to be centered around zero. This is what we expect for a line that fits the data reasonably well. Since the fit line should represent the overall pattern, we should expect some of the points to be above the fit line and some to be below it. That is, we should expect some of the residuals to be positive and others to be negative.

Example 12 *Looking at a set of residuals: 2*

Now let's look at the residuals when the line $y = 4.2 + 3.7x$ is fitted to the data on male unemployment and percentage of households without a car. The residuals from fitting this line are given in Table 7.

Table 7 Residuals from a different line fitted to the male unemployment and car ownership data

Town	x	y	Fit	Residual
Rutland	1.75	13.6	10.7	+ 2.9
Mole Valley	1.84	13.1	11.0	+ 2.1
West Dorset	2.14	16.9	12.1	+ 4.8
Bracknell Forest	2.25	14.5	12.5	+ 2.0
Rother	3.00	20.8	15.3	+ 5.5
Vale Royal	3.55	17.2	17.3	− 0.1
Alnwick	4.59	21.6	21.2	+ 0.4
Rotherham	5.19	29.7	23.4	+ 6.3
Dudley	5.27	25.3	23.7	+ 1.6
Norwich	5.61	35.5	25.0	+10.5

Notice that all but one of the residuals are positive. This pattern suggests that the fit values are generally too low. Hence, the line should be higher if it is going to fit the data reasonably well.

Example 13 *Looking at a set of residuals: 3*

Some data on oxygen uptake were introduced in Activity 6 (Subsection 2.1). A straight line was fitted to these data and the resulting residuals are given in Table 8. (In the table, only every fifth residual is given to make the table simpler.)

Table 8 Residuals from the oxygen-uptake experiment

Oxygen uptake	Residual
667	+ 4.7
1020	− 3.4
1599	−10.9
1874	−11.4
2312	−12.3
2766	− 6.8
3151	− 2.3
3521	+ 6.0
3878	+14.8
4290	+58.7

You can see that the residuals are positive for both small and large values of oxygen uptake (the explanatory variable) and they are negative for intermediate values.

This has a pattern to it. It is a pattern that is related to the values of the explanatory variable. This suggests that we could look for a relationship between the residuals and the values of the explanatory variable – and hence look for a relationship between the response variable and the explanatory variable which is over and above that explained by the fit line. So the fit line does not capture all of the relationship between the response variable and the explanatory variable.

The conclusion at the end of Example 13 holds in general.

Residual patterns

If the residuals show a pattern that relates to the explanatory variable, then the fit line does not provide an adequate explanation of all the patterns in the data, and we should look for a better relationship.

You have now covered the material related to Screencast 3 for Unit 5 (see the M140 website).

Some patterns in residuals are easy to spot by looking at a table. However, it is more usual to investigate a possible relationship between the residuals and the values of the explanatory variable using a scatterplot in which the horizontal coordinate is the explanatory variable, exactly as in the original scatterplot, and the vertical coordinate is the residual. Such a scatterplot is called a **residual plot** or sometimes a residual scatterplot.

Example 14 *A residual plot*

Figure 32(a) shows a scatterplot of the oxygen uptake and expired ventilation data from Figure 6, together with a fit line. (To make the plot simpler, only every fifth data point is shown – similarly to Table 8.) Figure 32(b) shows the corresponding residual plot. On the residual plot, notice a line corresponding to a residual value of zero – shown for reference.

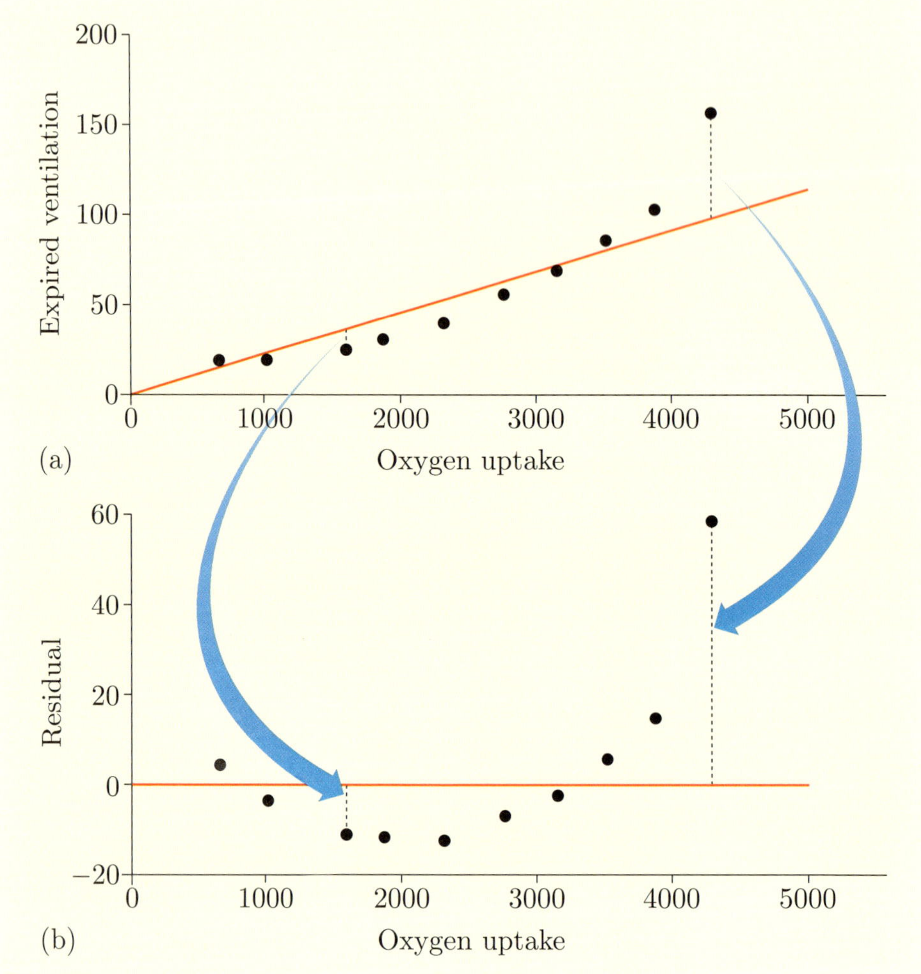

Figure 32 Data from an experiment in kinesiology, with (a) a straight fit line and (b) showing residuals

The residual plot shows a definite curved pattern. A straight line does not provide a satisfactory fit to the data. The residuals actually look larger in the residual plot than they do in the scatterplot, because the vertical scale has been increased to show the residuals more clearly. In this example, the curve could be seen in the original scatterplot, but often patterns are easier to spot in a residual plot.

Example 14 is the subject of Screencast 4 for Unit 5 (see the M140 website).

Activity 16 *Working with a residual plot*

An experiment was carried out at Charing Cross Hospital on the effect of the drug captopril (Figure 33) on the blood pressure of patients with moderate essential hypertension. The diastolic blood pressure of 15 patients was measured immediately before, and two hours after, receiving an injection of the drug. (Source: MacGregor, Markandu, Roulston and Jones (1979), *British Medical Journal*, vol. 2, pp. 1106–1109)

(Note that 'moderate essential hypertension' is a disorder involving blood pressure and 'diastolic blood pressure' is the lowest pressure between heartbeats.)

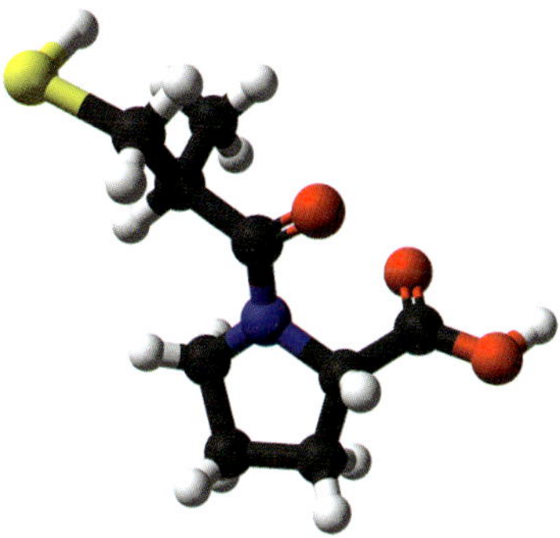

Figure 33 A model of the drug captopril

The results of the experiment are shown in the scatterplot in Figure 34 and a straight line has been fitted by eye to this data. (Obviously, blood pressure *before* treatment must be the explanatory variable, as this can influence blood pressure after treatment, but this cannot be true the other way round.)

Figure 35 shows four possible residual plots for these data and the line.

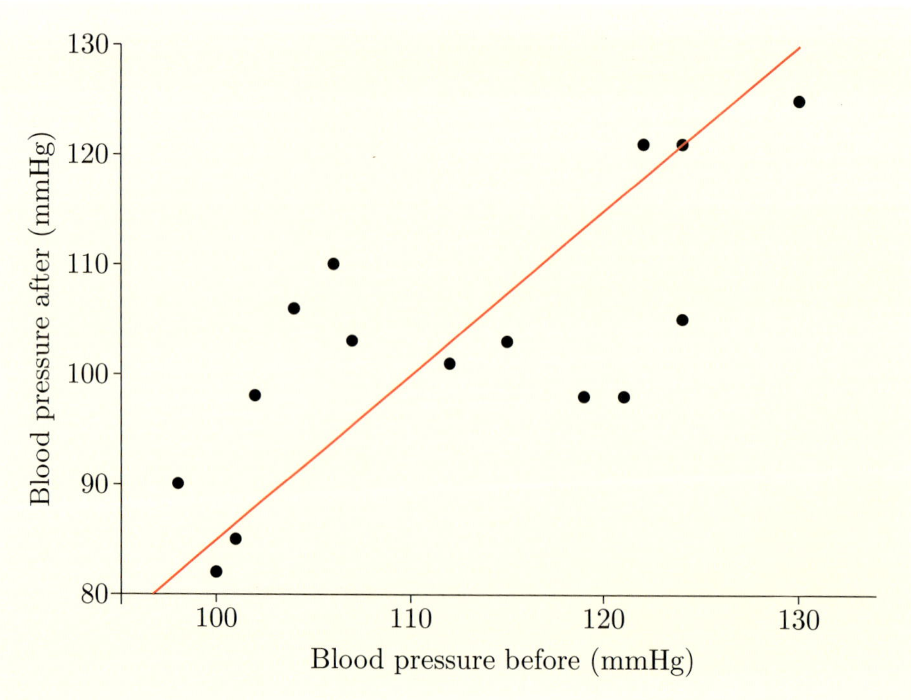

Figure 34 Blood pressure data from captopril study

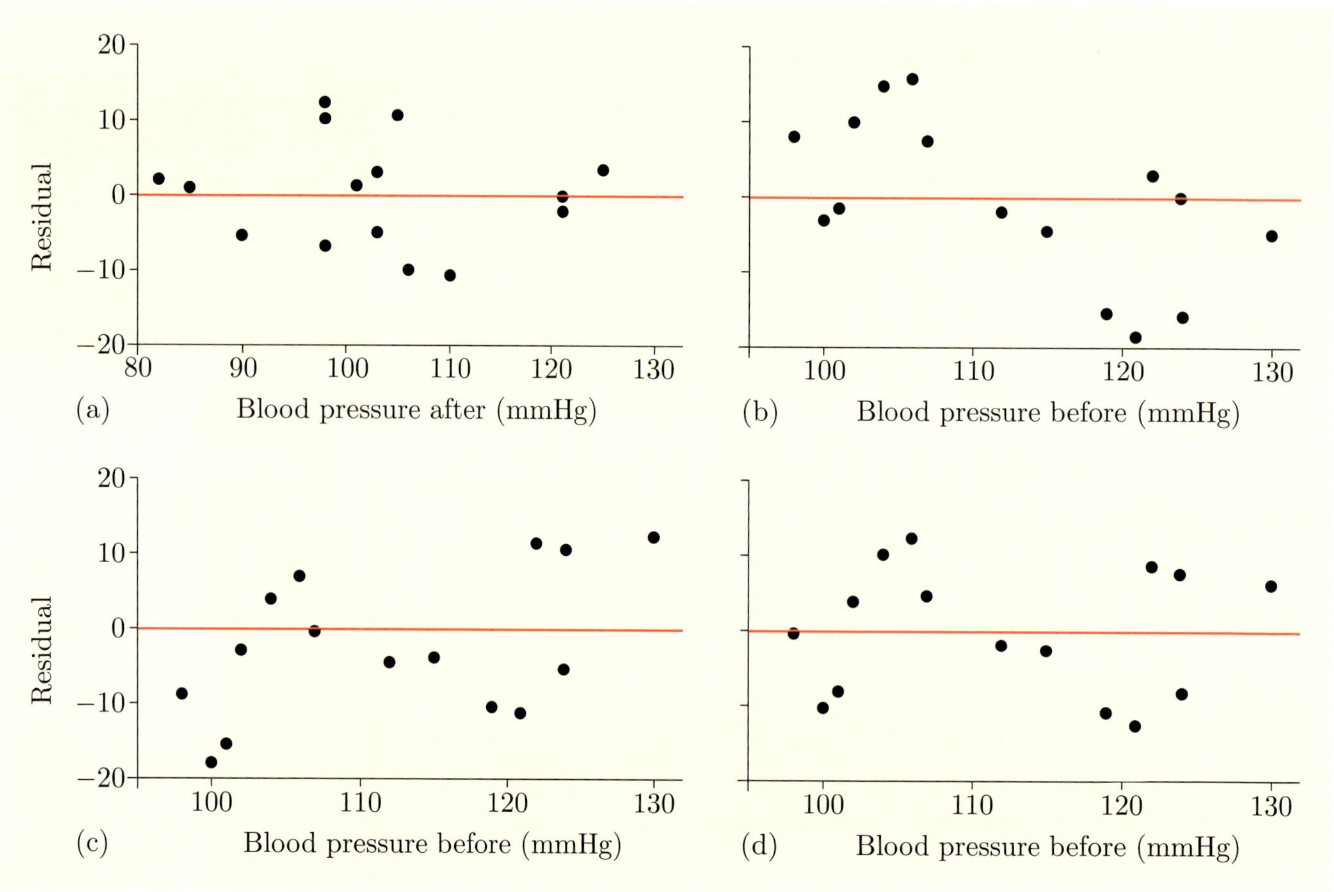

Figure 35 Four possible residual plots for the blood pressure data

(a) Which one of the four residual plots in Figure 35 corresponds to the correct residual plot for the line shown in Figure 34?

(b) Can you spot a pattern in the correct residual plot from Figure 35? If so, how should the fit line in Figure 34 be moved?

In Activity 16, the residuals suggest that the line fitted by eye is not the best that could be drawn and that a line which is a little less steep would be better. We could proceed by drawing another line, finding the residual values and drawing another residual plot. But this is rather a hit-and-miss procedure. We might find that it is now not steep enough, or perhaps that it was a little too high so that too many residuals were negative. Also, no two people would end up drawing exactly the same straight line. Moreover, the procedure is very tedious, particularly if there are a lot of points. In Section 4 you will learn a method of calculating the equation of a straight line that provides a good fit to a set of data.

Exercises on Section 3

Exercise 5 *Summarising change in house prices over time*

Exercise 4 featured a scatterplot of average prices in the UK between 1991 and 2008. This scatterplot is repeated below for convenience.

Use this scatterplot to add a line that you feel provides a good summary of the data.

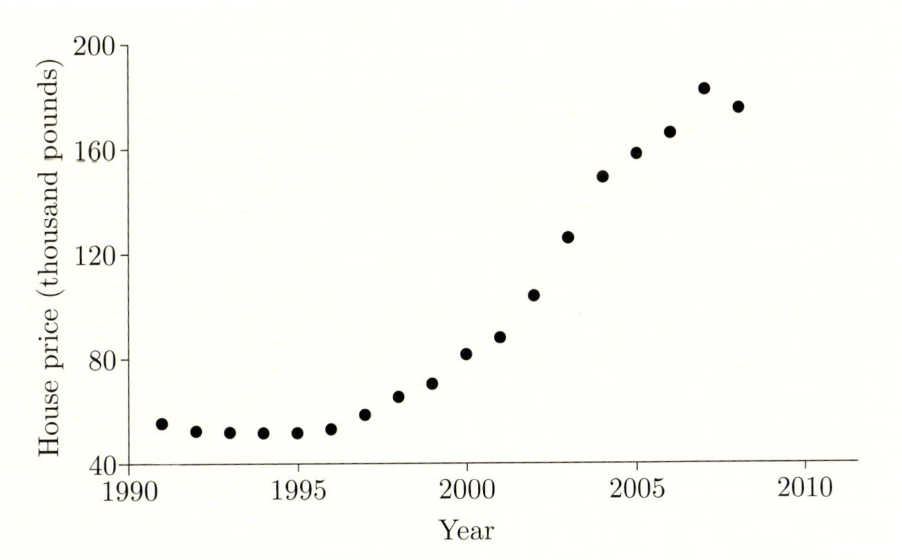

Figure 36 Average house price in the UK between 1991 and 2008

Exercise 6 *Calculating fitted values and residuals*

For a data point where $x = 20$ and $y = 4$, calculate the following.

(a) The fit value when the equation of the line is $y = 125 - 6x$.

(b) The fit value when the equation of the line is $y = -3 + 0.25x$.

(c) The residual when the equation of the line is $y = 0.15x$.

(d) The residual when the equation of the line is $y = 8 + x$.

Exercise 7 *Assessing the fit of lines*

Four different lines have been fitted to a set of 20 data points. The corresponding residual plots are shown below. Also shown on each plot is the line corresponding to a residual value of zero. Using each plot, comment on the fit of the line. If you think that the line does not fit very well, suggest how the line should be moved so that it fits the data better.

(a)

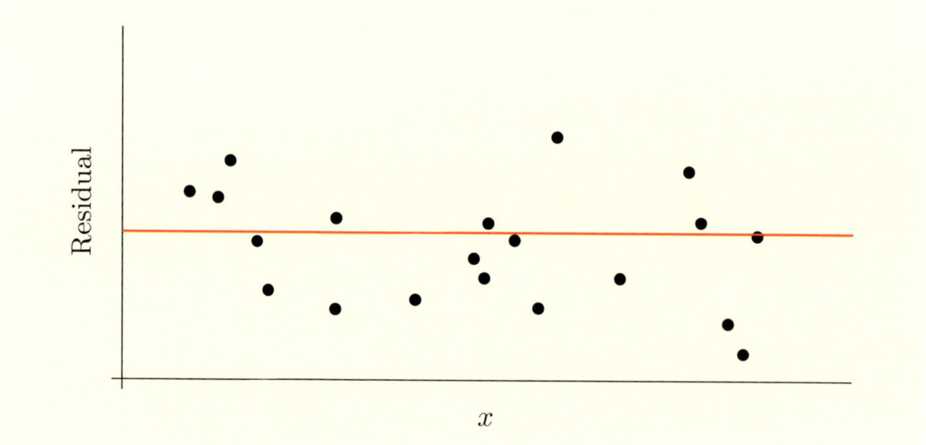

Figure 37

(b)

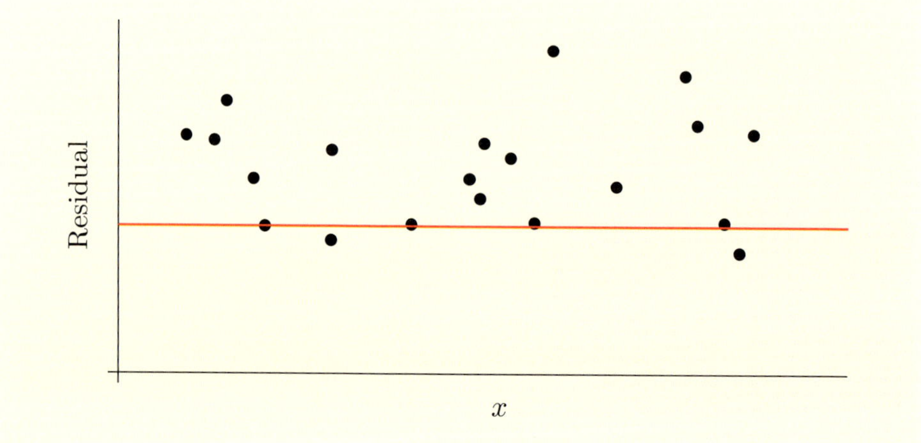

Figure 38

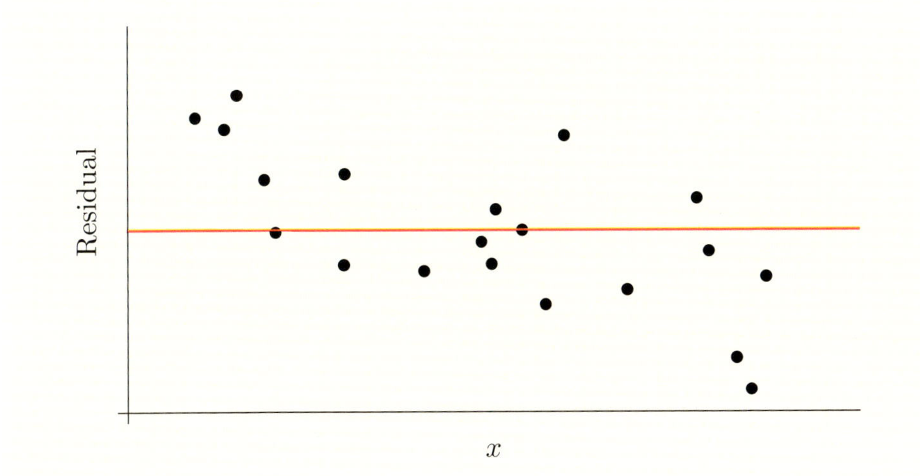

Figure 39

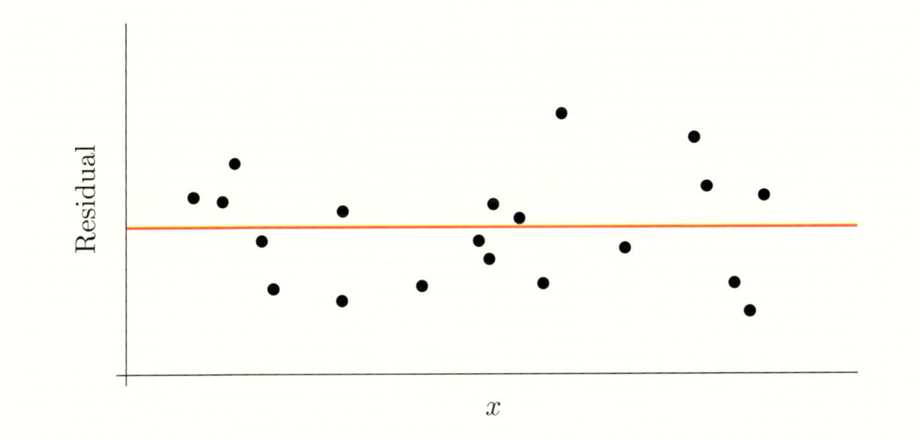

Figure 40

4 The least squares regression line

In the last section, we concentrated on fitting lines by eye. This is a simple, straightforward method and is often an adequate approach when exploring data to get some idea of what relationship is present. However, fitting lines by eye is subjective – different people draw different lines. If you are going to report your investigation to someone else, there is a need for an objective, well-defined procedure for drawing a straight fit line on a scatterplot. If you have used a formal method, you can describe exactly what you did, and another person presented with the same dataset would get exactly the same line by using your method.

Another reason for requiring a formal method is that when computers are to be used to fit lines, they have to be instructed exactly how to carry this out; a computer cannot just draw a line that 'appears' to be a good fit! Also, it is good to choose a straight line that is optimal in some way.

Several formal methods of fitting lines exist. In this section, we introduce what is by far the commonest of these, which is known as fitting by the method of **least squares**. It has many useful properties, some of which will be discussed here. The resulting line is called the **least squares fit line** or the **least squares regression line**.

4.1 What is least squares?

The method of least squares is based on a study of residuals obtained when different lines are fitted to a set of data.

A line is a good fit to a batch of data if the residuals are small. When all of the points lie exactly on a straight line, as in Example 4 (Subsection 2.2), all the residuals are zero and the fit is perfect. However, this is very rarely the case in practice, and so we need a method that chooses a line for which the residuals are as small as possible.

In Example 12 (Subsection 3.3) all but one of the residuals were positive and we suggested that the line was too low. Instinctively, a good line should be somewhere in the middle of the data. The method of least squares takes care of this.

The method of least squares

This method is used to find a good fit line, by choosing a line that passes through the overall mean of the data: the line goes through the point whose x-coordinate is the mean of the x-values in the data and whose y-coordinate is the mean of the y-values in the data. This point can be denoted $(\overline{x}, \overline{y})$, where $\overline{x}$ and $\overline{y}$ are the two means.

It can be shown that if a line passes through the point $(\overline{x}, \overline{y})$, then the sum of all the residual values, taking their signs into account, is always zero. In other words, the total of all the positive residuals is equal to the total of all the negative residuals. You do not have to worry about why this is so, but you can see it is a useful property of the fit line; it ensures that there are not too many positive or too many negative residuals. It also takes account of the situation where we might have one or two large positive residuals and all the other residuals are small and negative. It is unnecessary for there to be equal numbers of positive and negative residuals.

Requiring the fit line to go through $(\overline{x}, \overline{y})$ only gives one point that the line must pass through; something else is needed to choose the best slope. Look at Figure 41.

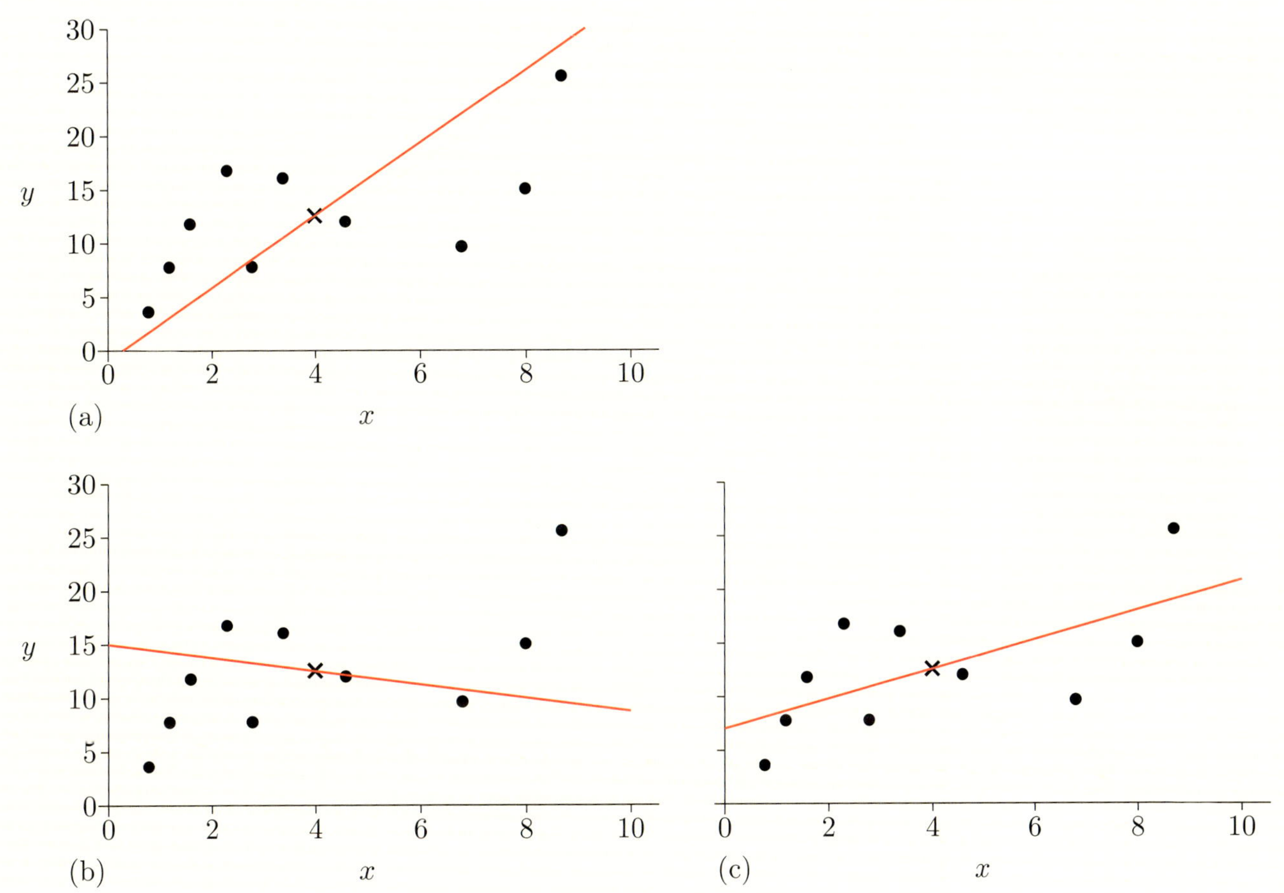

Figure 41

This figure shows three possible fit lines for the same set of data points. All three lines pass through the point $(\overline{x}, \overline{y})$, which is marked with a cross. The line in Figure 41(a) is too steep, and the line in Figure 41(b) is not steep enough. In both of these, the lengths of some of the residuals, ignoring their signs, are quite large. By contrast, the lengths of all the residuals in Figure 41(c) are small, and this line is the best fit of the three lines illustrated.

Residual plots for the lines shown in Figure 41(b) and (c) have been drawn in Figure 42(a) and (b). Look just at the *lengths* of the residuals. You can see that for almost every point the residual is shorter in Figure 42(b) than in (a). This is what you expect: the line in Figure 41(c) looks to be a better fit than the line in Figure 41(b) *because* it gives shorter residuals.

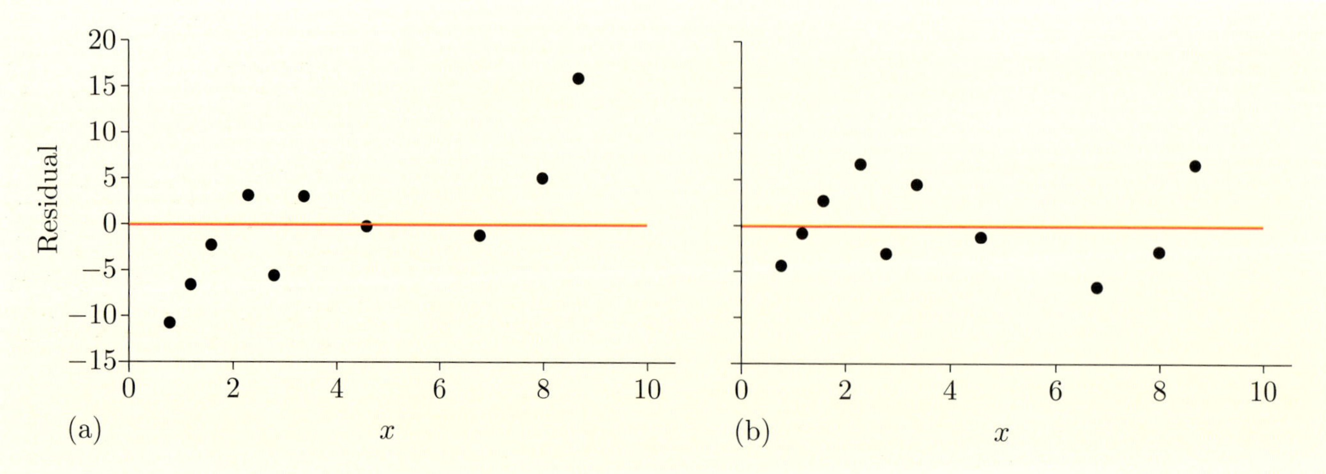

Figure 42 Residual plots corresponding to Figure 41(b) and (c)

The method of least squares uses the fact that lines that give a good fit have short residuals. To get rid of the negative signs, the lengths of all residuals are squared. These squares are all added together, and the slope is chosen so as to make this sum of squared residuals as small as possible. It is this property that gives the method its name – *least squares*.

The least squares regression line

The least squares regression line is the line for which the sum of the squares of the residuals is minimised.

The least squares regression line always goes through the point $(\overline{x}, \overline{y})$ and hence the sum of the residuals is always zero.

Sir Francis Galton and regression

Sir Francis Galton (1822–1911) made important contributions in many fields. He was obsessed with data and measured everything that he could, from wind direction (he was an initiator of scientific meteorology) to fingerprints. The latter led to him devising a method of classifying fingerprints that proved useful in forensic science. He was also a geographer and explorer in his early years (notably in south-west Africa) and founded the science of measuring mental faculties (psychometrics). However, he is best known for his work in anthropology, heredity and eugenics, and statistics.

He was a half-cousin of Charles Darwin and was greatly influenced by Darwin's book *Origin of Species*, published in 1859. Following its publication, much of Galton's work focused on exploring variation in human populations and on whether human ability was inherited or learned – he coined the phrase *nature or nurture.* To better understand the large quantities of data that he collected, he devised and extended a number of statistical techniques, including regression.

One dataset he collected were the heights of 205 parents and their 930 adult children. The heights of women were multiplied by 1.08, so as to adjust for gender. Galton then plotted the height of a child (the response) against the average height of its parents (the explanatory variable) and represented their relationship by a straight line. Galton noted that the children of the shorter parents tended to be taller than their parents, and the children of the taller parents tended to be shorter than their parents. (Source: Galton, F. (1886) 'Regression towards mediocrity in hereditary stature', *The Journal of Anthropological Institute of Great Britain and Northern Ireland*, vol. 15, pp. 246–263.)

This regression of a child's measurement towards the mean value in the population is a characteristic of inherited attributes. It is called *regression to the mean.* This example underlies the use of *regression* in the phrase *least squares regression.*

Sir Francis Galton

You have now covered the material needed for Subsection 5.1 of the Computer Book.

4.2 Calculating the least squares regression line by hand

Calculating the least squares regression line bears many similarities to calculating the standard deviation for a single variable, in particular the calculation using Method 2. So you may find it helpful to revise Subsection 3.1 of Unit 3 before studying this subsection.

The least squares regression line has the form

$$y = a + bx.$$

We must calculate its slope, b, and its intercept, a. It turns out that the formula for the slope of the least squares regression line is as follows:

The summation notation was first introduced in Subsection 1.3 of Unit 2.

$$b = \frac{\sum(y - \overline{y})(x - \overline{x})}{\sum(x - \overline{x})^2}$$

and that the formula for the intercept of the least squares regression line is

$$a = \overline{y} - b \times \overline{x}.$$

These formulas look daunting at first, and how they are derived from the definition of the least squares regression line is beyond the scope of M140. So the method for calculating the regression line using least squares will be demonstrated in this subsection by means of an example. We will calculate the regression line for the data on percentages of males unemployed and households with no car.

The calculation will be broken down into five steps. The first step is to calculate the sum of all the x-values, the sum of all the y-values, the sum of the squares of all the x-values and the sum of the products of the x- and y-values. That is, we will calculate

$$\sum x, \sum y, \sum x^2 \text{ and } \sum xy.$$

Example 15 *Calculating a least squares regression line – step 1*

In the scatterplots of these data in this unit, the percentage of men unemployed plotted along the x-axis, and the percentage of households with no car along the y-axis. So in terms of x and y, the data are as follows.

Table 9 Male unemployment and car ownership for ten towns in England

Town	x	y
Alnwick	4.59	21.6
Vale Royal	3.55	17.2
Rotherham	5.19	29.7
Rutland	1.75	13.6
Dudley	5.27	25.3
Norwich	5.61	35.5
Bracknell Forest	2.25	14.5
Rother	3.00	20.8
Mole Valley	1.84	13.1
West Dorset	2.14	16.9

The sums of all the x-values and of all the y-values are as follows:

$$\sum x = 4.59 + \cdots + 2.14 = 35.19, \qquad \sum y = 21.6 + \cdots + 16.9 = 208.2.$$

We also require the sum of the squares of the x-values and the sum of the products of the x- and y-values. You should be able to find these two sums on your calculator without writing down each square (or product) separately.

$$\begin{aligned}\sum x^2 &= 4.59^2 + 3.55^2 + \cdots + 2.14^2 \\ &= 144.9419,\end{aligned}$$

$$\begin{aligned}\sum xy &= 4.59 \times 21.6 + 3.55 \times 17.2 + \cdots + 2.14 \times 16.9 \\ &= 825.928.\end{aligned}$$

These four sums are the basic quantities you need, and this completes the first step of the calculations.

The second step is to calculate the means of the x- and y-values. The third step is to calculate the sum of the squared deviations of the x-values and the sum of the products of the deviations of the x- and y-values. That is, in steps 2 and 3, we will calculate

$$\overline{x},\ \overline{y},\ \sum (x - \overline{x})^2 \text{ and } \sum (x - \overline{x})(y - \overline{y}).$$

Example 16 *Calculating a least squares regression line – steps 2 and 3*

Step 2 is the calculation of means of the x and y values.

In this dataset there are ten observations, so $n = 10$.

The mean of x is therefore $\overline{x} = 35.19/10 = 3.519$, and the mean of y is therefore $\overline{y} = 208.2/10 = 20.82$.

In step 3, the sum of the squared deviations of the x-values is calculated. This sum is one that you have encountered before, as part of the

The calculation of the standard deviation was introduced in Subsection 3.1 of Unit 3.

calculation of a standard deviation. Here part of 'Method 2' for the standard deviation is used to calculate $\sum(x-\overline{x})^2$.

$$\begin{aligned}\sum(x-\overline{x})^2 &= \sum x^2 - \frac{(\sum x)^2}{n}\\ &= 144.9419 - \frac{(35.19)^2}{10}\\ &= 144.9419 - 123.833\,61 = 21.108\,29.\end{aligned}$$

The sum of the products of the deviations of the x- and y-values can be calculated in a similar way using the sum of the products of the x- and y-values along with the mean of the x-values and the mean of the y-values.

$$\begin{aligned}\sum(x-\overline{x})(y-\overline{y}) &= \sum xy - \frac{(\sum x)(\sum y)}{n}\\ &= 825.928 - \frac{35.19 \times 208.2}{10}\\ &= 825.928 - 732.6558 = 93.2722.\end{aligned}$$

The final two steps involve calculating the slope and the intercept of the regression line. The only terms required are those calculated in steps 2 and 3.

Example 17 *Calculating a least squares regression line – steps 4 and 5*

The slope, b, of the regression line is given by the following formula.

$$b = \frac{\sum(x-\overline{x})(y-\overline{y})}{\sum(x-\overline{x})^2}.$$

So in this example

$$b = \frac{93.2722}{21.108\,29} \simeq 4.418\,747,$$

which we round to three significant figures as 4.42 (the same number of significant figures as the percentage of men unemployed is given to).

The intercept, a, of the regression line is given by:

$$a = \overline{y} - b \times \overline{x}.$$

So in this example

$$a \simeq 20.82 - 4.418\,747 \times 3.519 = 5.270\,429\,307,$$

which we round to two decimal places as 5.27, one more decimal place than that used for the percentage of households without a car.

The equation of the least squares regression line is therefore

$$y = 5.27 + 4.42x.$$

Often the least squares regression line is referred to as simply 'the regression line', and calculating and using the line is referred to as **linear regression**. The procedure for its calculation can be summarised as follows.

Calculating the least squares regression line $y = a + bx$ for a set of n data points (x, y)

1. Calculate $\sum x$, $\sum y$, $\sum x^2$ and $\sum xy$.
2. Calculate the means of x and y:
$$\overline{x} = \frac{\sum x}{n} \quad \text{and} \quad \overline{y} = \frac{\sum y}{n}.$$
3. Calculate the sum of the squared deviations of the x-values
$$\sum (x - \overline{x})^2 = \sum x^2 - \frac{\sum x^2}{n},$$
and the sum of the products of the deviations
$$\sum (x - \overline{x})(y - \overline{y}) = \sum xy - \frac{\sum x \sum y}{n}.$$
4. The slope b is given by
$$b = \frac{\sum (x - \overline{x})(y - \overline{y})}{\sum (x - \overline{x})^2}.$$
5. The intercept a is given by
$$a = \overline{y} - b\overline{x}.$$

You have now covered the material related to Screencast 5 for Unit 5 (see the M140 website).

To draw the fitted line on a scatterplot, we just calculate the coordinates of two well-separated points on the line and then draw a straight line through them. After drawing the line you should look at it to check that it seems right – it should appear to pass through the middle of the data. If it clearly does not, then there is a calculation error and you should check your working.

Example 18 *Drawing the regression line on a scatterplot*

In Example 17, the regression line was calculated as

$$y = 5.27 + 4.42x.$$

From Table 9 in Example 15, the scatterplot, the x-values of the data range from about 2 to 6. We substitute these values into the equation of the regression line to obtain the coordinates of two well-separated points.

When $x = 2$,

$$y = 5.27 + 4.42 \times 2 = 14.11,$$

so one point on the line is (2, 14.11).

When $x = 6$,

$$y = 5.27 + 4.42 \times 6 = 31.79,$$

so a second point on the line is (6, 31.79).

Figure 43 shows the line drawn on the scatterplot. You can see that it appears to provide a reasonably good fit to the points. If you compare it to Figure 23 (Subsection 3.1), you can see that the line fitted by least squares is slightly steeper than the one drawn by eye.

Figure 43 Percentage of males unemployed and percentage of households with no car, with least squares line

The next activity provides practice in calculating a regression line.

Activity 17 *Calculating a least squares regression line*

Activity 16 (Subsection 3.3) featured a line fitted by eye to diastolic blood pressure data from a study of the drug captopril. The data shown in Figure 34 of Activity 16 are given in Table 10. Using these data, calculate the least squares regression line. Also, add the regression line to Figure 44.

Table 10 Diastolic blood pressure before and after injection

x, blood pressure before injection (mmHg)	y, blood pressure after injection (mmHg)
130	125
122	121
124	121
104	106
112	101
101	85
121	98
124	105
115	103
102	98
98	90
119	98
106	110
107	103
100	82

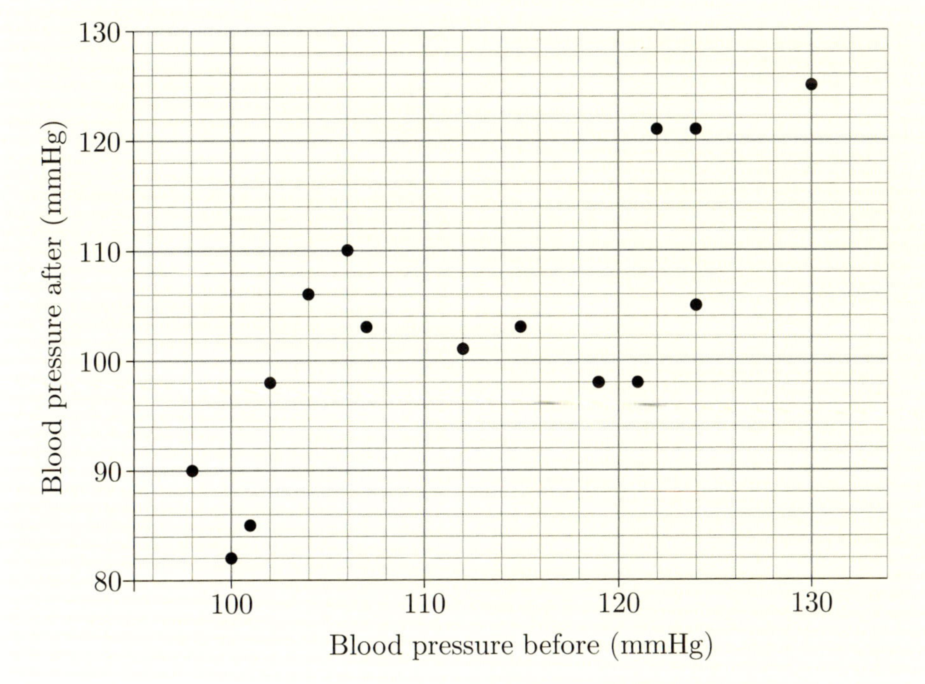

Figure 44 Scatterplot of blood pressure data from captopril study

Exercises on Section 4

Exercise 8 *Spotting the least squares regression line*

In Figure 45, three lines fitted to a set of 12 data points are shown. One of these lines is the least squares regression line. Identify which one it is. For each of the two lines that are not the least squares regression line, give a reason why it is not.

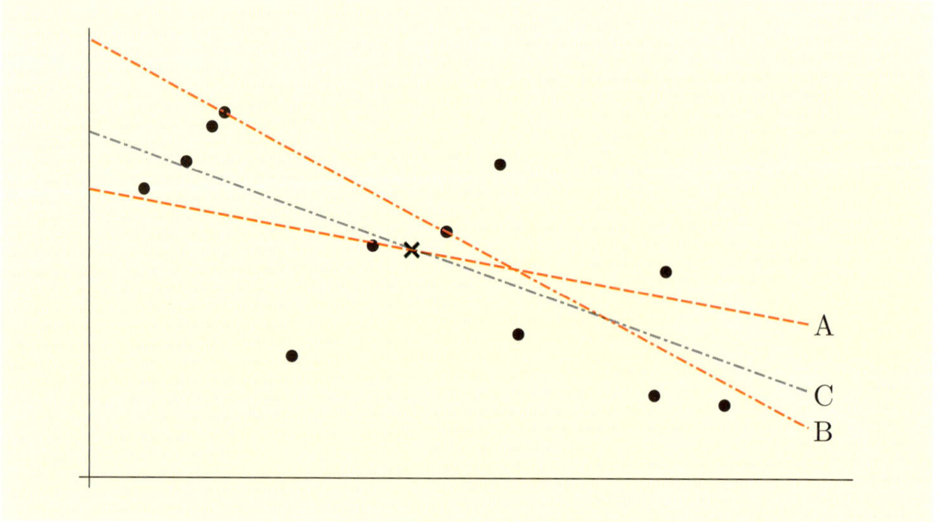

Figure 45 Scatterplot of some data along with three fitted lines. (The point $(\overline{x}, \overline{y})$ is also marked on the scatterplot.)

Exercise 9 *Fitting a line to house prices*

In the 1980s, the average UK house prices were as follows. Using these data, calculate the least squares regression line.

Table 11 Average house prices in the UK

x, Year	y, House price (£ thousands)
1980	23.3
1981	24.1
1982	24.7
1983	27.4
1984	30.8
1985	34.2
1986	37.0
1987	43.0
1988	48.9
1989	62.2

(Data source: Nationwide building society (2013) 'UK house prices since 1952')

5 Using the least squares regression line

The previous section introduced the least squares regression line and showed how it can be calculated. Once we have calculated a regression line for a sample of data points, what can we do with it? In this section we will explore two things: checking that the regression line fits the data well, and using the line for prediction.

'Data don't make any sense, we will have to resort to statistics.'

5.1 Checking the least squares regression line

In Section 6 you will see that least squares is a method by which a straight line can be fitted to data automatically. Although the least squares regression line should be as good as any straight line fitted by eye, it is possible that no straight line fits the data well. So, it is important to check that the least squares line provides an adequate fit. This can be done by looking at the residuals.

Since we have used the least squares method to fit the line, there is no need to look for two of the patterns that we sometimes found in Subsection 3.3. The average of the residuals is zero, so there cannot be a pattern of too many positive (or negative) residuals. It is possible that there will be more, say, small positive residuals and fewer large negative residuals, but the least squares method chooses a line where the sum of the residuals is zero. It also chooses the slope so that the sum of the squared residuals is as small as possible. This ensures that there will be no tendency for positive residuals to be associated with, say, small values of x, and negative residuals with large values of x. Otherwise, the line would be rotated about the point $(\overline{x}, \overline{y})$, as we saw in Figure 41.

The patterns that we *might* see include, for example, positive residuals for both small and large values of x, and negative residuals for intermediate x, which indicates that a curved line would fit the data better.

Activity 18 *Interpreting residual plots*

The following three residual plots are the result of fitting least squares lines to three different sets of data. Use each residual plot to state how reasonable a straight-line model is for the dataset. Justify your opinion.

(a)

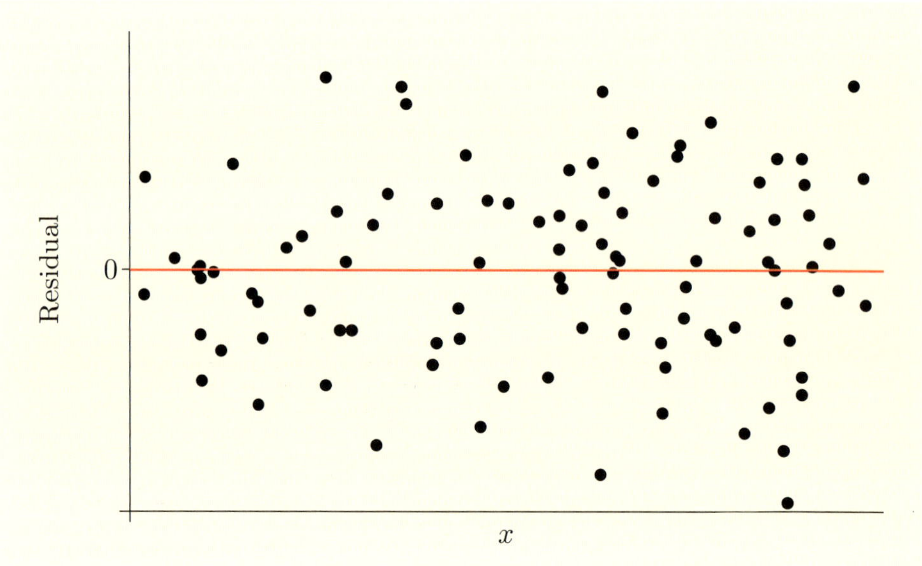

Figure 46 Residual plot for dataset 1

(b)

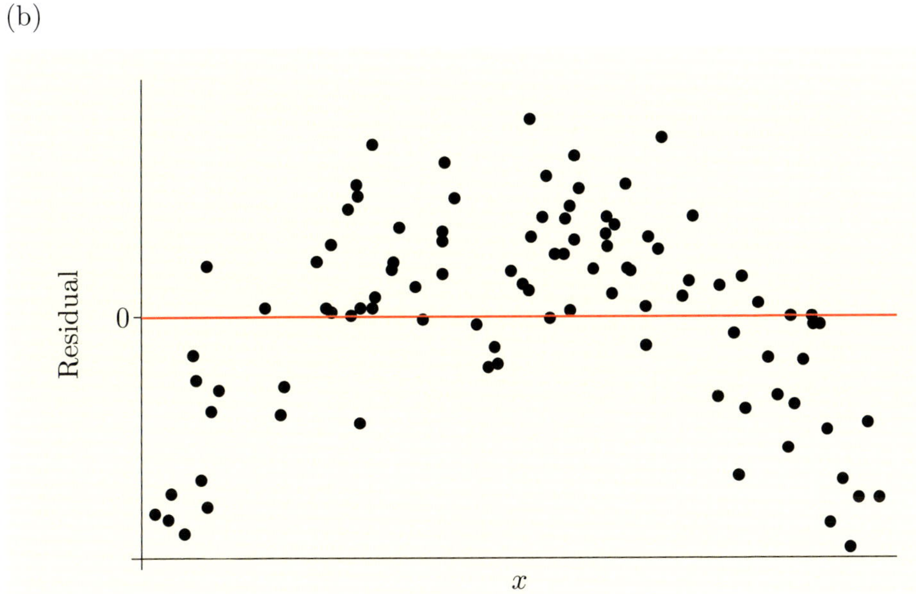

Figure 47 Residual plot for dataset 2

(c)

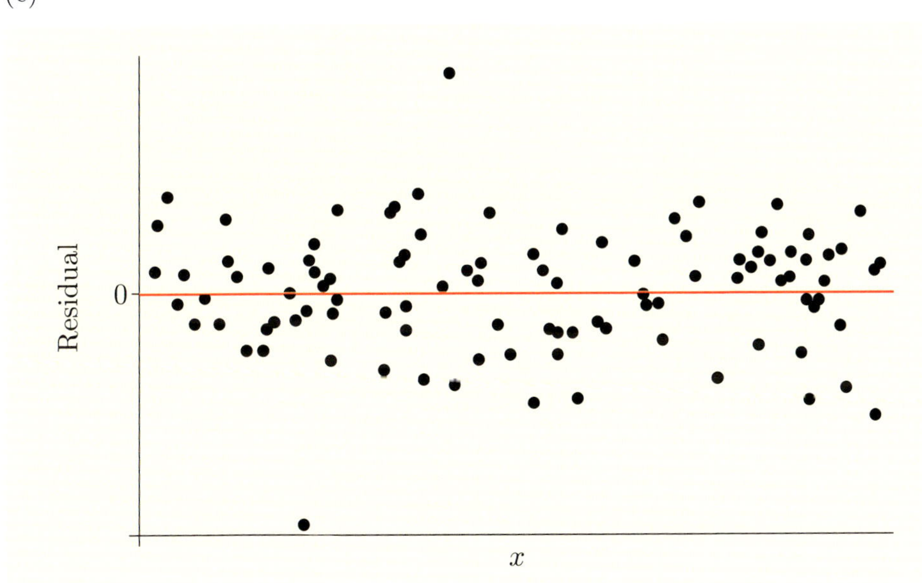

Figure 48 Residual plot for dataset 3

So, when the line fits well there is no pattern in the residual plot. That is the residuals are evenly scattered above and below the line $y = 0$. Possible patterns that indicate problems with the fit include the following:

- A non-linear relationship between residuals and the explanatory variable (which indicates that a curved line will probably be more appropriate).
- A particularly large or small residual. That is, one that does not follow the pattern of the other residuals. This indicates that there is probably an outlier in the data.

Residuals can be found most easily using the equation of the least squares line, following the same process as in Subsection 3.2. If the line has the equation $y = a + bx$, then for each value of x in the sample, we calculate the fit value and then the residual.

$$\text{Fit} = a + bx.$$

$$\begin{aligned}\text{Residual} &= \text{Data} - \text{Fit}\\ &= y - (a + bx).\end{aligned}$$

Let us return to the example on male unemployment and households without cars.

Example 19 *Examining residuals for a least squares regression line*

The least squares regression line calculated in Subsection 4.2 was $y = 5.27 + 4.42x$. We shall use this line to calculate the residual for the first data point, Alnwick, (4.59, 21.6):

$$\begin{aligned}\text{Fit} &= 5.27 + 4.42 \times 4.59\\ &= 5.27 + 20.2878\\ &= 25.6 \text{ (rounded to one decimal place).}\end{aligned}$$

So,

$$\begin{aligned}\text{Residual } &\simeq 21.6 - 25.6\\ &= -4.0.\end{aligned}$$

Table 12 shows all the fit and residual values (rounded to one decimal place).

Table 12 Residuals and fitted values for ten towns

Town	x	y	Fit	Residual
Alnwick	4.59	21.6	25.6	−4.0
Vale Royal	3.55	17.2	21.0	−3.8
Rotherham	5.19	29.7	28.2	+1.5
Rutland	1.75	13.6	13.0	+0.6
Dudley	5.27	25.3	28.6	−3.3
Norwich	5.61	35.5	30.1	+5.4
Bracknell Forest	2.25	14.5	15.2	−0.7
Rother	3.00	20.8	18.5	+2.3
Mole Valley	1.84	13.1	13.4	−0.3
West Dorset	2.14	16.9	14.7	+2.2

Figure 49 shows the residual plot for this example. There is no obvious pattern to be seen here, so this suggests that a straight line is a reasonable model for the data points.

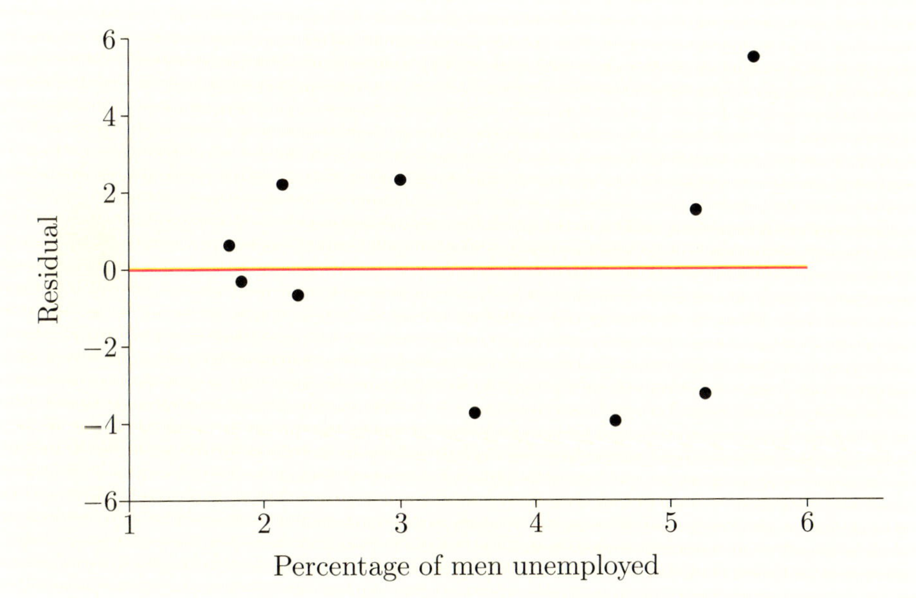

Figure 49 Residual plot for the male unemployment and car ownership data, using the least squares line

Example 19 is the subject of Screencast 6 for Unit 5 (see the M140 website).

Activity 19 *Examining the fit of a least squares regression line*

In Activity 17 (Subsection 4.2) you calculated the least squares regression line for the blood pressure data from the study involving the drug captopril.

(a) Using the equation of the regression line, $y = 4.2 + 0.880x$, calculate the residuals for the first five observations. For convenience, the data for these observations are given again below.

x	y
130	125
122	121
124	121
104	106
112	101

(b) Figure 50 shows the residual plot for all the data. Comment on the fit of the regression line to these data.

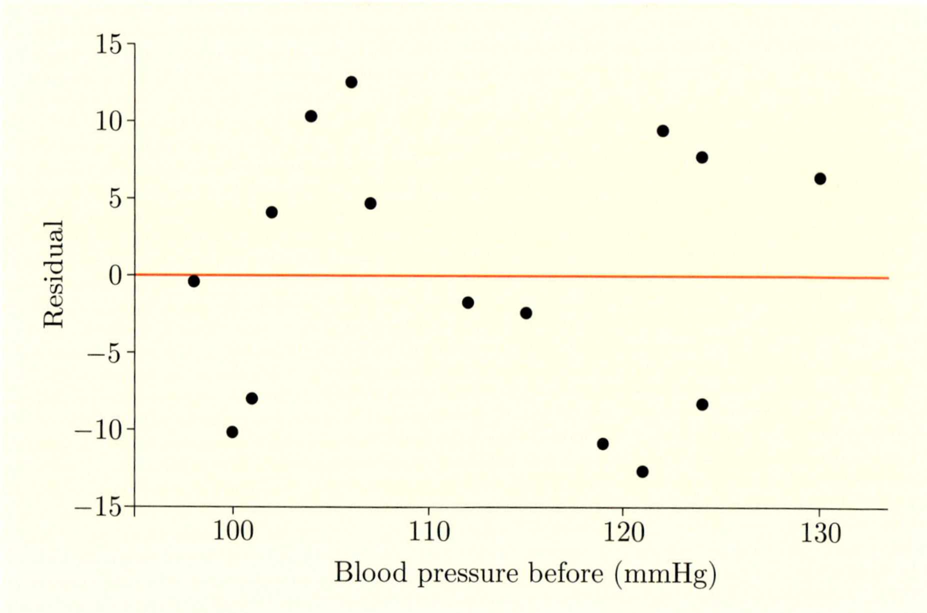

Figure 50 Residual plot for the blood pressure data, using least squares regression

5.2 Using the least squares regression line for prediction

Once we have established that the regression line does provide a reasonable model for a set of data, what can we do with it? One of its uses is for **prediction**. If we know the value of the explanatory variable for some individual, then we can forecast the value of the response variable for that individual.

'Mr Palmer, using statistics, I can predict which numbers will be chosen in the lottery .... I just don't know when.'

In Activity 17 (Subsection 4.2), you calculated the following regression line for diastolic blood pressure before (x) and after (y) injection with captopril for patients with moderate essential hypertension: $y = 4.2 + 0.880x$. Suppose another patient with moderate essential hypertension arrives at the hospital. If a doctor measures the patient's diastolic blood pressure, then by using the equation of the regression, the doctor can calculate the fit value. This value is a prediction of the patient's diastolic blood pressure two hours after injection with captopril. For example, if the patient's blood pressure on arrival was 124 mmHg, then the doctor would expect a blood pressure of $4.2\,\text{mmHg} + 0.880 \times 124\,\text{mmHg} \simeq 113\,\text{mmHg}$ after treatment.

Activity 20 *Predicting values*

(a) Suppose the doctor measures another patient's diastolic blood pressure on arrival and found it to be 105 mmHg. Using the regression line found in Activity 17, predict the patient's blood pressure two hours after injection with captopril.

(b) Suppose a town was found to have 3.78% of men unemployed in the 2001 census. Using the regression line, $y = 5.27 + 4.42x$, that was used in Example 19 (Subsection 5.1), predict the percentage of households in that town that had no car.

It is important to note that a regression line can only be used to predict the response, y, from the explanatory variable, x. It cannot be used to predict x from y. This is because the response variable and explanatory variable are treated differently when we calculate the equation of a regression line. Least squares minimises the square of the *vertical* distances from the points to the line, not the square of the horizontal distances. Minimising the squared vertical distances on a scatterplot and minimising the squared horizontal distances lead to different 'best' lines.

Suppose you were told that 22.0% of households in a particular town did not have a car, and you were asked to predict the town's male unemployment rate. The regression equation used in part (b) of Activity 20 is of no help – it was calculated with a town's male unemployment rate as the *explanatory* variable. To predict the town's male unemployment rate, a new regression line must be calculated, with male unemployment rate as the *response* variable, and the percentage of households with no car as the explanatory variable. To highlight the new roles of the two variables, we could use y^* to denote the percentage of men unemployed (the response variable) and x^* to denote the percentage of households with no car (the explanatory variable). Using the data in Table 3 (Subsection 1.2), the least squares regression line with y^* as the response is

For comparison, the equation of the line $y = 5.27 + 4.42x$ can be rearranged as $x = -1.19 + 0.226y$.

$$y^* = -0.403 + 0.188x^*.$$

So we predict that in 2001 the town would have a male unemployment rate of $-0.403 + 0.188 \times 22.0 \simeq 3.73$.

Prediction should only be done from explanatory variable to response.

Let us think a bit about exactly what is meant by prediction. In the first place, the prediction is an average. We cannot possibly say that a patient with blood pressure of 124 mmHg will necessarily find it is reduced to exactly 113 mmHg by the treatment. In fact, if you look at the original sample, you see that two of the patients did have initial blood pressure of 124 mmHg; one dropped to 121 mmHg and the other to 105 mmHg after treatment. As we saw in Subsection 3.1, a line summarising a relationship usually does not go through any of the data points. If the relationship is

All the data are given in Table 10, Activity 17 (Subsection 4.2).

strong, then there is little scatter of the points about the straight line, and so we can expect the actual value of the response variable to be close to the predicted one. For the blood pressure example, there is a moderate amount of scatter about the line, so that is not the case. (The scatterplot of the blood pressure data, along with the least squares regression line, is given in the solution to Activity 17.) Hence the predicted value only gives a rough indication of the response variable.

There is another way of considering the predicted value: as an average. If a lot of patients with blood pressure of 124 mmHg are treated in the same way, then the least squares line tells us that their average blood pressure after treatment will be about 113 mmHg.

The fitted value $y = a + bx$ is an estimate of the average value of the response variable Y that occurs when the explanatory variable takes the value $X = x$.

Activity 21 *Interpreting predictions*

(a) For the blood pressure example, the predicted value when $x = 110$ is 101. Interpret this predicted value.

(b) Using the least squares regression line for the data on male unemployment (the explanatory variable) and households with no car (the response variable), the predicted value when $x = 4.00$ is 22.95. Interpret this predicted value.

5.3 Applicability of the least squares regression line

An important point which applies to all the discussions on relationships in this unit is that conclusions only apply to the populations from which the original data were taken. The following two examples examine the applicability of the fitted lines found for two sets of data.

Example 20 *Applicability of a fitted line: 1*

In Activity 16 (Subsection 3.3), we were told that the data are measurements of blood pressure on patients who were suffering from moderate essential hypertension. In the absence of any further information, we can assume that it was a random sample of patients with this complaint. The least squares line you calculated in Activity 17 is only appropriate to such patients. Patients suffering from high blood pressure for a different reason might react quite differently to the drug. It might have been the case that all patients were women between the ages of 25 and 40. Then it would not be valid to use the line to make a prediction about a man or a 60-year-old woman suffering from moderate essential hypertension. They might react differently from young women.

Example 21 *Applicability of a fitted line: 2*

In Examples 15 to 17 (Subsection 4.2), we fitted a least squares line to data relating to male unemployment and car ownership. These data were from a sample of towns and small regions of England. London and other major cities were excluded. Again, this would have to be stressed in any conclusions. Results would not necessarily apply to the city of Birmingham or to a town in Scotland.

Furthermore, the data used were all from the 2001 census. The regression line may not be appropriate for data from a different census. Thus it may not be valid to use this line to make a prediction about the percentage of households without a car at the time of the 2011 census.

A final point to be aware of when considering prediction from any regression line is that a prediction is only valid for the range of values of x, the explanatory variable, represented in the original sample. For the blood pressure example, all the patients in the initial sample had initial blood pressure between 98 and 130. What about a patient with initial blood pressure of 150? This is so far outside the original range that we do not know what the scatterplot would be like there. Perhaps the straight-line model would no longer apply. The drug might be more effective or less effective for a patient with exceptionally high blood pressure. It is only reasonable to predict for an x-value within or perhaps a little outside the range of values of x in the original sample.

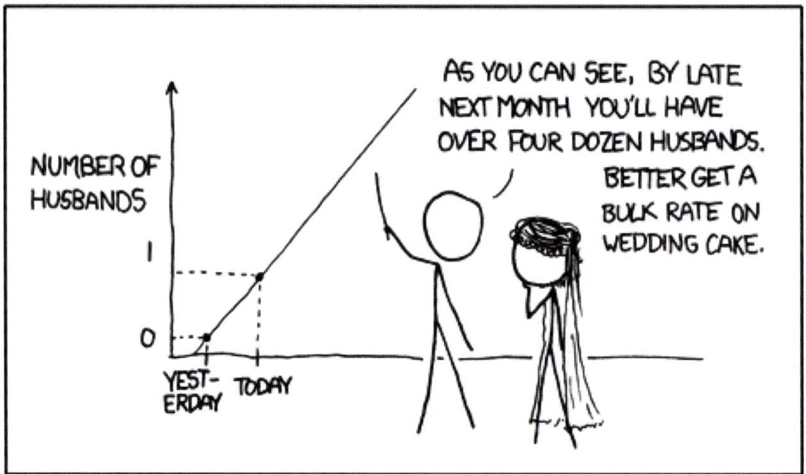

Activity 22 *Assessing the reasonableness of predictions*

Activity 8 (Subsection 2.3) introduced data on student achievement in different countries. The data plotted in Figure 15, reproduced below, is based on the performance of 15-year-olds in the different countries in 2009.

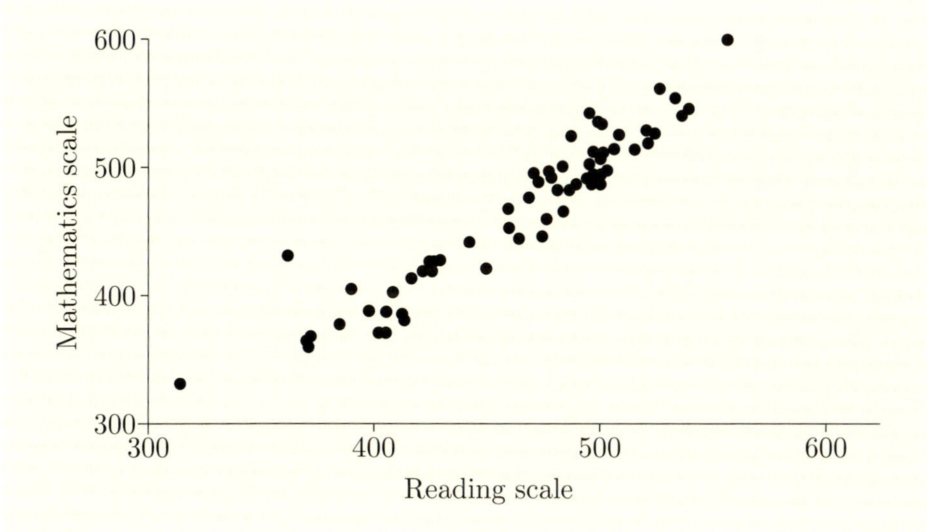

Figure 51 Student performance in reading and mathematics

The equation of the least squares line fitted to the data is $y = 42.7 + 1.099x$. For each of the following, state how reasonable you think it is to use this fitted line to predict the average performance of 15-year-olds of a country on the mathematics scale in 2009.

(a) When the performance of 15-year-olds on the reading scale in 2009 is 400.

(b) When the performance of 12-year-olds on the reading scale in 2009 is 400.

(c) When the performance of 15-year-olds on the reading scale in 2009 is 200.

(d) When the performance of 15-year-olds on the reading scale in 2009 is 575.

Exercises on Section 5

Exercise 10 *Matching residual plots*

Match the following plots of data and least squares regression line with the corresponding residual plots. In each case state whether the regression line provides a reasonable summary of the relationship between x and y.

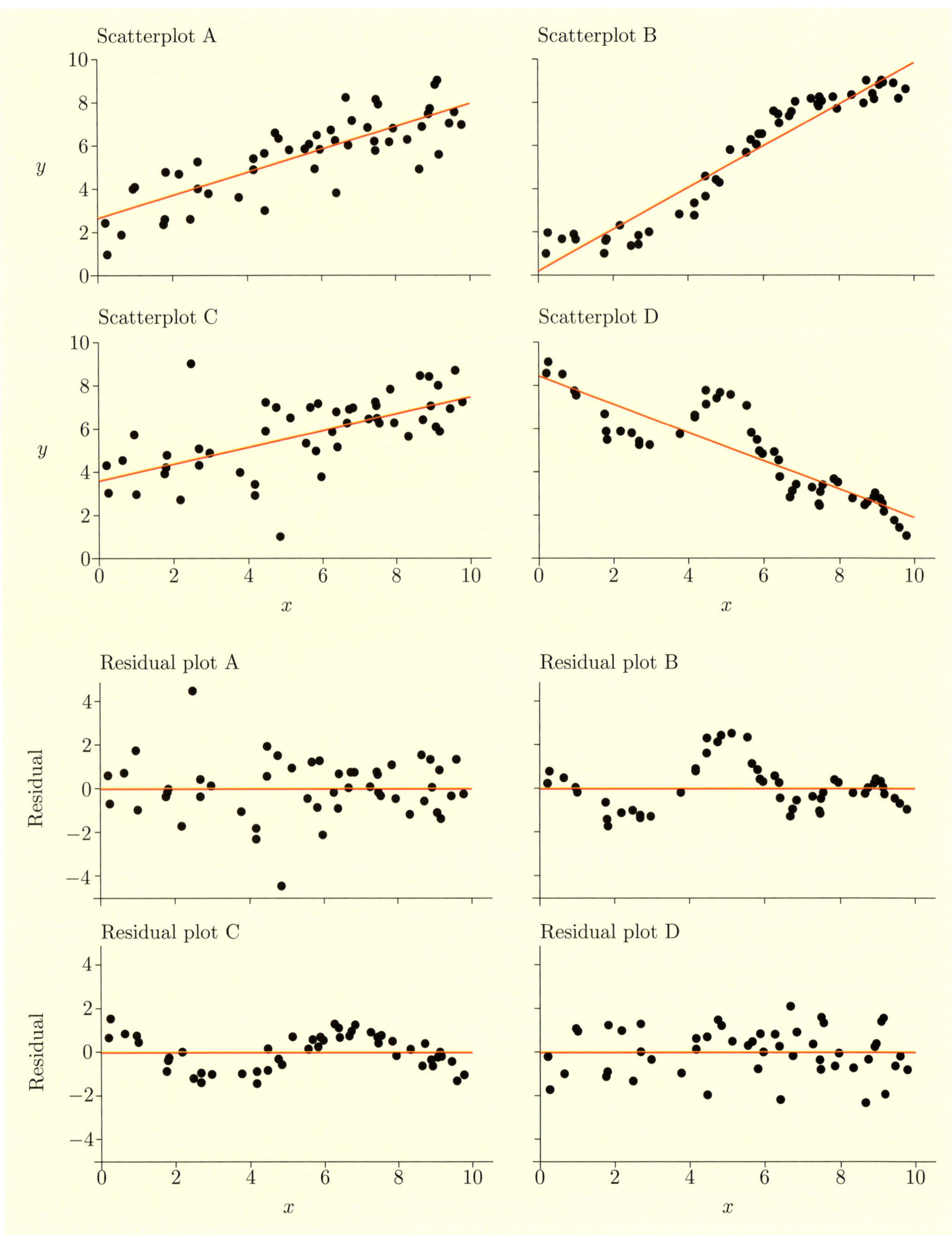
Scatterplot A
Scatterplot B
Scatterplot C
Scatterplot D
y
x
0
2
4
6
8
10
Residual plot A
Residual plot B
Residual plot C
Residual plot D
Residual
−4
−2
0
2
4

Exercise 11 *Predicting house prices*

In Exercise 4 some data on average house prices in the UK between 1991 and 2008 were introduced.

A least squares regression line turns out to have the equation

$$\text{price (in thousands)} = -16\,992.0 + 8.546 \times \text{year}.$$

(a) Use this line to predict average house prices in 2009, 2010 and 2030. (Give your answer rounded to the nearest £1000.)

(b) Comment on the reasonableness of the predictions you calculated in part (a).

(c) The corresponding residual plot for the regression line is shown in Figure 52. Use it to comment on whether a straight-line model is suitable for these data. Does this change the conclusion you came to in part (b). If so, in what way?

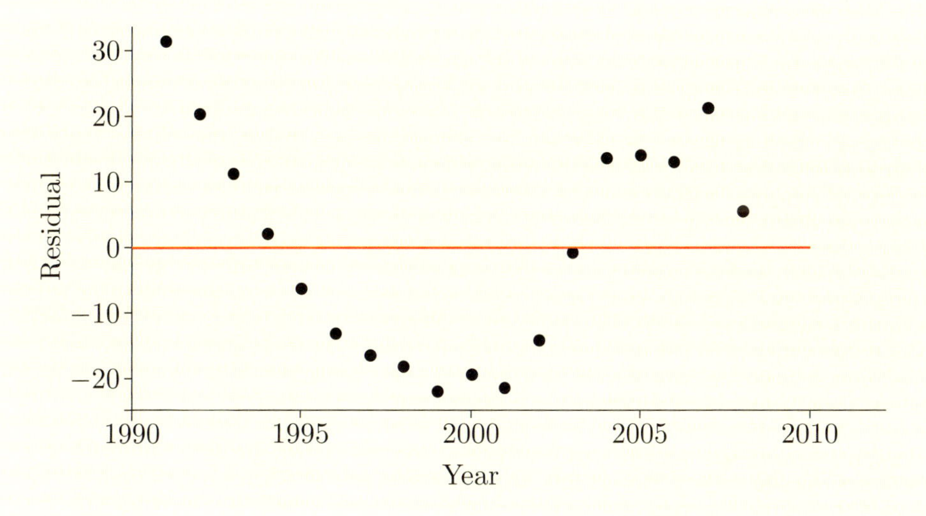

Figure 52 Residual plot for average house prices in the UK using the least squares regression line

6 Computer work: relationships

So far in this unit, you have been calculating the least squares regression line and the corresponding residuals by hand. In this section, you will learn how to do the following using Minitab:

- obtain the least squares regression line
- obtain a scatterplot with the regression line displayed on it
- obtain residual plots.

You should now turn to the Computer Book and work through Subsection 5.1, if you have not already done so, followed by the rest of Chapter 5.

Summary

The theme for this unit has been relationships between linked variables. You have learned how to investigate relationships by looking at scatterplots. That is, to assess whether a relationship appears to be: positive, negative or neither; linear or non-linear; and strong, weak or not present. You also learned that outliers on scatterplots are points that do not appear to follow the same pattern as the other points.

Relationships are summarised by drawing on a scatterplot the simplest adequate line that represents the pattern of points. In many cases, this simplest adequate line is just a straight line. Straight lines can be fitted to data 'by eye', that is subjectively choosing the line, or by using least squares regression, where the line is found that minimises the sum of the squared residuals. You learned how to calculate the equation of the least squares line $y = a + bx$ by hand and by using Minitab. Regression lines can be can be used to predict values of the response variable.

Finally, you have learned that the fit of a regression line can be investigated by calculating residuals. A pattern in the residuals suggests that the line does not capture all of the relationship, so the line does not fit the data well. A residual plot is often used to help spot any patterns in the residuals. You learned how to produce residual plots using Minitab.

Learning outcomes

After working through this unit, you should be able to:

- explain what is meant by a relationship between two variables
- understand the terms response variable and explanatory variable, and decide which is which in a given example
- recognise positive and negative relationships from a scatterplot
- explain what is meant by two variables being linearly related, and recognise this from a scatterplot
- describe a relationship between two variables which is neither positive nor negative
- recognise strong and weak relationships from a scatterplot
- recognise outliers in a scatterplot
- draw a straight line by eye to fit a scatterplot
- find the residuals from a straight fit line
- recognise patterns in a residual plot
- understand what is meant by 'least squares' in the context of fitting lines to data
- calculate a least squares regression line for a batch of linked data by hand and by using Minitab
- produce a residual plot from a scatterplot and the least squares regression line using Minitab
- use a regression line to predict the value of the response variable, and know when it is appropriate to do this.

Solutions to activities

Solution to Activity 1

(a) You might find some children of different ages and measure their heights. It would be best to choose a random sample of children. This would ensure that you did not select particularly tall or especially short children without realising it. It would be a good idea to choose separate samples of boys and girls, as their heights at the same age might follow a different pattern.

(b) There are many ways in which you could describe the relationship numerically. One possibility is something like 'boys grow about 10 centimetres a year on average from age 6 to age 12'. You might have suggested a more complicated relationship, describing the different rates of growth at different ages, or you might have suggested a completely different relationship.

(c) No, the heights of adults do not generally vary with age.

Solution to Activity 2

No, these figures do not tell us anything about such a relationship. We are told unemployment rates only for Bedfordshire regions, and the percentage of households with no car only for Merseyside regions. We need to know both figures for each of the regions to find out about the relationship. In other words, we need *linked* data.

Solution to Activity 3

(a) No, these data are not linked. This is because the measurements of year-7 heights are not measured on the same children as the year-6 heights.

(b) Yes, these are linked data. The data consist of 20 pairs of measurements, each pair being the height of a single child both one year ago and now.

Solution to Activity 4

(a) According to Table 3, in Vale Royal there were 3.55% of males unemployed and 17.2% of households with no car. So the coordinates for Vale Royal are $(3.55, 17.2)$.

Similarly, the coordinates for Rother are $(3.00, 20.8)$.

(b) The point in the top rightmost corner of Figure 2 lies at the point corresponding to about 5.5 along the x-axis, and at the point corresponding to about 35 along the y-axis. So the coordinates for this point are roughly $(5.5, 35)$. Looking at Table 3, Norwich has 5.61% of males unemployed and 35.5% of households with no car. No other town in the list has approximately 5.5% males unemployed and roughly 35% of households with no car. So this point must correspond to Norwich.

(c) A low male unemployment rate tends to be associated with a low percentage of households with no car, and a high male unemployment rate tends to be associated with a higher percentage of households with no car. However, the relationship is not exact. For example, if about 17% of households in a town have no car, the unemployment rate could be as low as 2.1% (West Dorset) or as high as 3.6% (Vale Royal), but it is unlikely to be as high as 5.5%.

Solution to Activity 5

(a) In this case, the amount of fertiliser will probably affect the yield of the tomato plant, but the yield cannot affect the amount of fertiliser. The amount of fertiliser is chosen by the experimenter. So the amount of fertiliser is the explanatory variable and would be plotted on the x-axis. Yield would be plotted on the y-axis.

(b) In this case, the percentage of households with no car should be the response variable, and the percentage of males unemployed the explanatory variable. If a man is unemployed, it is reasonable to assume that household income is usually lower and so the household is less likely to be able to afford a car. If, on the other hand, a household does not have a car, this would not normally cause a man to lose his job. So this means that the percentage of males unemployed should be plotted on the x-axis of the scatterplot, as was done in Figure 2 (Subsection 1.3).

(However, you may have felt that if a household does not have a car, it may limit the job opportunities available to members of that household. So that as the percentage of households without a car goes up, the more likely it is for men to be unemployed. In this case it is the percentage of households with no car that should be plotted on the x-axis.)

(c) This is a situation where the choice is not clear-cut. There is no particular reason to say that the consumption of water by metered households of a water company depends on the consumption of water by its unmetered households. Equally, there is no particular reason to think that consumption of water by unmetered households depends on the consumption by the metered households. Hence either quantity could be plotted on the x-axis. (But deciding that you wanted to predict one of these quantities from the other would change this.)

Solution to Activity 6

(a) Adding a shaded area that encloses all the points gives the following scatterplot. Notice that the shaded area slopes downwards from left to right, so the two variables are negatively related.

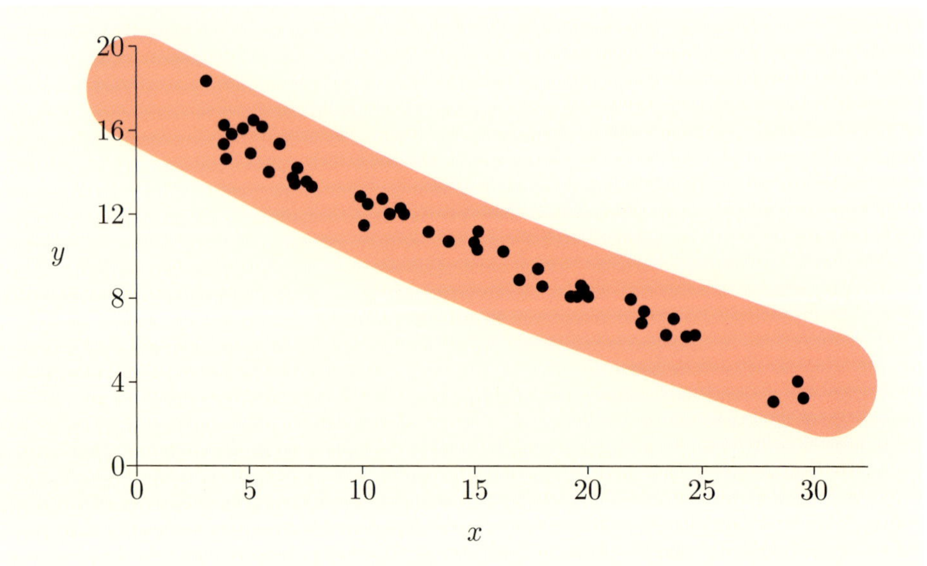

(b) The scatterplot for the kinesiology experiment is given below. Notice that in this case the shaded area enclosing all the points is curved, not straight. However, this makes no difference as to whether the relationship is positive or negative. The shaded area goes up from left to right. That is, low values of x tend to be associated with low values of y, and high values of x with high values of y. So there is a positive relationship between oxygen uptake and expired ventilation.

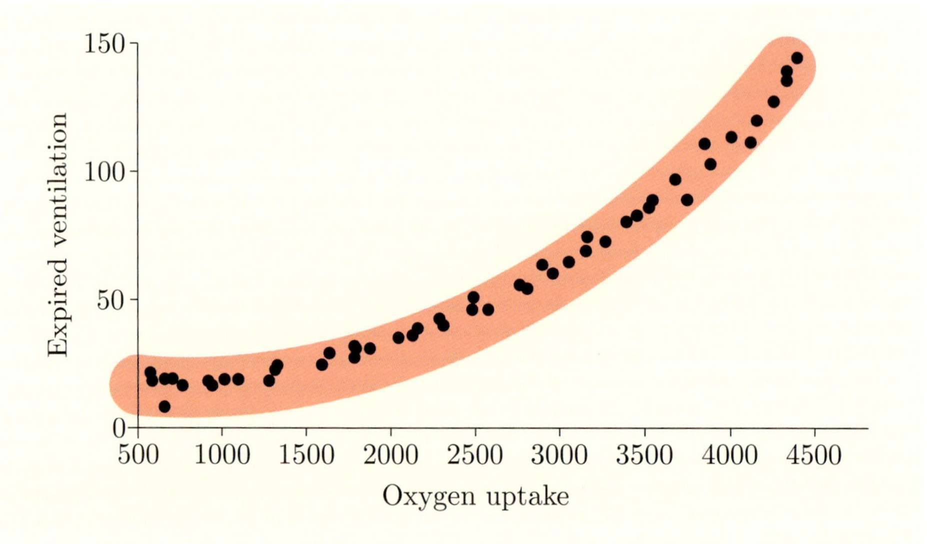

Solution to Activity 7

A straight line would not adequately follow the general pattern of points. Any reasonable line needs to be curved. So the relationship between the metal distance and the ultrasonic response is non-linear – a negative non-linear relationship, to be more precise.

Solution to Activity 8

From strongest to weakest: Figure 16, then Figure 15 and, lastly, Figure 14.

Figure 16 shows the strongest relationship. The points in this scatterplot all lie close to the general pattern.

The points in Figure 15 also lie close to the general pattern, just not as close as in Figure 16. So the relationship in Figure 15 is not as strong as that in Figure 16.

In contrast, the points in Figure 14 do not lie close to any general pattern, so much so that a general pattern is hard to pick out. So the relationship in Figure 14 is quite weak, certainly weaker than those in Figures 15 and 16.

Solution to Activity 9

There is no discernible pattern in the points. If there is any relationship between these variables, it is very weak.

Solution to Activity 10

(a) In Figure 19 there is one obvious outlier (ringed below). It is a country whose value on the mathematics scale is much higher than would be expected given its value on the reading scale. Another country has an extremely low score on the reading scale. However, it also has a low score on the mathematics scale that seems in keeping with its score on the reading scale (given the other data), so it would not be considered an outlier.

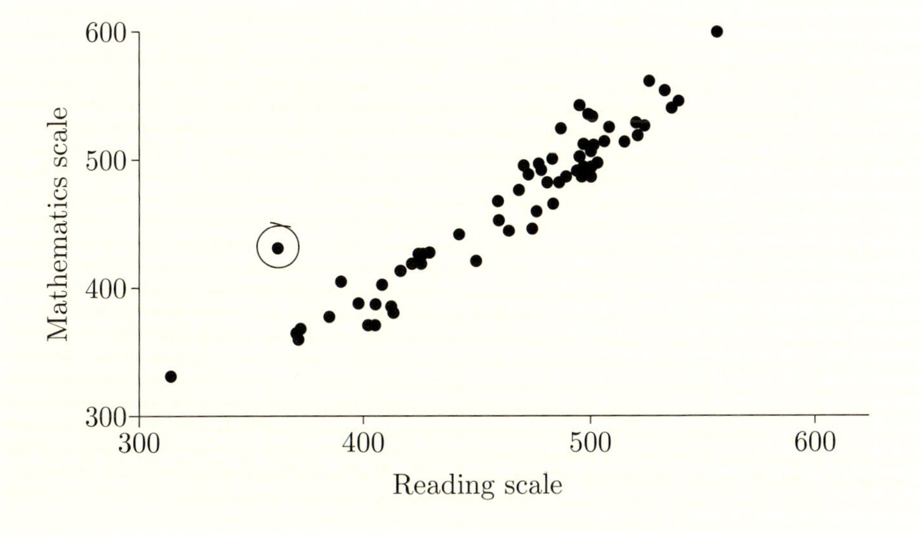

(b) The same data were plotted in Figure 4 (Subsection 2.1) along with a shaded area indicating a general pattern. All the points on Figure 4 lie within the shaded area, indicating that all the points fit with the general pattern. So there do not appear to be any outliers. (The point with coordinates (400, 13.2) looks a little high, but not so high as to make it an outlier.)

Solution to Activity 11

(a) Temperature is the explanatory (independent) variable, and yield is the response variable. This is reasonable, as the yield cannot explain temperature, which is chosen by the experimenter.

(b) Temperature and yield appear to have a reasonably strong non-linear relationship. This relationship is not clearly positive or negative, as it goes up and then down. There do not appear to be any outliers.

(c) The scatterplot with one version of a summary line is shown in the following figure. Notice that this line matches the description of the relationship given in part (b). It is a curve, not a straight line; it goes up and then down as you move from left to right; and all of the points lie close to the curve.

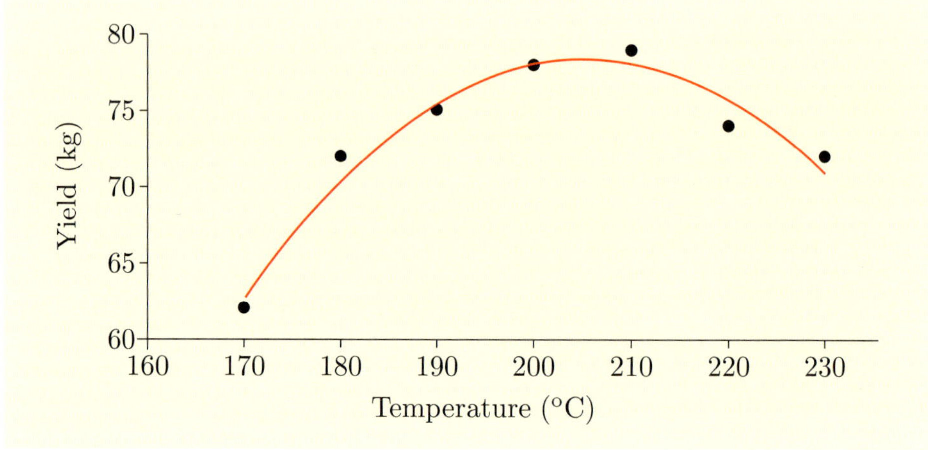

Yield from an industrial process, with one possible summary line

Solution to Activity 12

The line in Figure 27(a) is obviously not a good choice. It is too high up, with only two points on it and none above it. It should be moved down.

The line in Figure 27(b) is also not a good choice. It is too steep and does not go anywhere near the three points in the top right corner or the three points in the bottom left. It should be made less steep.

The line in Figure 27(c) is a much better fit than the lines in Figure 27(a) and (b), as it is fairly close to the points. There are more points below it than above it, though, so it might be better a little lower.

The line in Figure 27(d) also appears to be quite a good choice. It might be better if it were a little steeper.

Solution to Activity 13

The fit values are read off the scatterplot, and the residuals calculated as $y - \text{Fit}$. On the scatterplot, the 'Residual' is the vertical distance from the data point to the line.

Region	x	y	Fit	Residual
England				
North East	372.7	12.0	12.4	−0.4
North West	430.5	11.4	11.5	−0.1
Yorkshire and the Humber	405.5	11.4	11.9	−0.5
East Midlands	449.4	11.6	11.3	+0.3
West Midlands	430.1	11.7	11.5	+0.2
East	493.4	10.9	10.6	+0.3
London	577.8	9.7	9.3	+0.4
South East	523.8	10.6	10.1	+0.5
South West	482.6	11.3	10.8	+0.5
Wales	394.0	13.1	12.1	+1.0
Scotland	447.2	11.4	11.3	+0.1
Northern Ireland	482.8	11.8	10.8	+1.0

For the second point in Table 4, corresponding to the North West, $x = 430.5$. If you draw a vertical line through this point, it meets the fitted line at $y = 11.5$. So the fit value is 11.5. Hence the residual value is

$$\text{Data} - \text{Fit} = 11.4 - 11.5 = -0.1.$$

The other values are found in a similar way.

Since the y-values are given to one decimal place, it is good practice to read the fit values and so calculate the residual values to the same level of accuracy. It is in any case not possible to read the fit values from the graph any more accurately than this.

Solution to Activity 14

(a) $\text{Fit} = 2 + 4 \times 12 = 2 + 48 = 50.$

(b) $\text{Fit} = -4.6 + 0.3 \times 3 = -4.6 + 0.9 = -3.7.$

(c) $\text{Fit} = -0.5 \times (-2.5) = 1.25.$

(d) $\text{Fit} = -3.16 - 4.2 \times (-2.7) = -3.16 + 11.34 = 8.18.$

Solution to Activity 15

For the first point, Alnwick, $x = 4.59$ and the 'Data' (y) value is 21.6. The fit value is

$$5.8 + 4.2 \times 4.59 = 5.8 + 19.278 = 25.1$$

rounded to one decimal place. So the residual is

$$\text{Data} - \text{Fit} = 21.6 - 25.1 = -3.5.$$

The values for the rest of the points are calculated in a similar way. The completed table is as follows.

	x	y	Fit	Residual
Alnwick	4.59	21.6	25.1	−3.5
Vale Royal	3.55	17.2	20.7	−3.5
Rotherham	5.19	29.7	27.6	+2.1
Rutland	1.75	13.6	13.2	+0.4
Dudley	5.27	25.3	27.9	−2.6
Norwich	5.61	35.5	29.4	+6.1
Bracknell Forest	2.25	14.5	15.3	−0.8
Rother	3.00	20.8	18.4	+2.4
Mole Valley	1.84	13.1	13.5	−0.4
West Dorset	2.14	16.9	14.8	+2.1

Notice that it is still appropriate to give the fit values only to the same level of accuracy as the y-values. This means that the residual values also should be given to the same level of accuracy as the y-values.

Solution to Activity 16

(a) The correct residual plot is shown in Figure 35(b).

Figure 35(a) cannot be the corresponding residual plot because the residuals are plotted against the response variable (blood pressure after treatment) instead of the explanatory variable.

In Figure 34, notice that for the patients with the lowest blood pressure before treatment, two of them lie below the fitted line and one lies above the line. So on the residual plot, two of the three left-most points are negative and the other is positive. This only happens in Figure 35(b). The pattern of the other points above and below the fitted line in Figure 34 also only matches that in Figure 35(b).

(b) Looking at Figure 35(b), we can see that there is a tendency for positive residuals to occur with low values of blood pressure and negative residuals to occur with high blood pressure. It is not a very clear-cut effect – there are one or two points with the opposite sign in each case, but these exceptions have small values. The pattern would disappear if the fit line were rotated a little to make it a little less steep, as the residuals associated with low blood pressure would decrease and the residuals associated with high blood pressure would increase. There is no reason to move the fit line up or down, as overall the positive and negative residuals appear to be balanced.

Solution to Activity 17

1. The four initial sums required are as follows.

$$\sum x = 1685, \quad \sum y = 1546, \quad \sum x^2 = 190\,817, \quad \sum xy = 175\,019.$$

2. The mean of the x-values and the mean of the y-values are

$$\overline{x} = 1685/15 \simeq 112.333\,333\,3$$

and

$$\overline{y} = 1546/15 \simeq 103.066\,666\,7.$$

3. The sum of the squared deviations of the x-values is

$$\begin{aligned}\sum(x-\overline{x})^2 &= 190\,817 - \frac{(1685)^2}{15}\\ &\simeq 190\,817 - 189\,281.6667\\ &= 1535.3333,\end{aligned}$$

and the sum of the products of the deviations of the x- and y-values is

$$\begin{aligned}\sum(x-\overline{x})(y-\overline{y}) &= 175\,019 - \frac{1685 \times 1546}{15}\\ &\simeq 175\,019 - 173\,667.3333\\ &= 1351.6667.\end{aligned}$$

4. We can now calculate the slope, b, of the regression line:

$$\begin{aligned}b = \frac{\sum(x-\overline{x})(y-\overline{y})}{\sum(x-\overline{x})^2} &\simeq \frac{1351.6667}{1535.3333}\\ &\simeq 0.880\,373\,466.\end{aligned}$$

Don't worry, you won't have to join these chaps for long!

5. The intercept, a, of the regression line is then:

$$\begin{aligned}a &= \overline{y} - b \times \overline{x}\\ &\simeq 103.066\,666\,7 - (0.880\,373\,466 \times 112.333\,333\,3)\\ &\simeq 103.066\,666\,7 - 98.895\,285\,98 = 4.171\,380\,72.\end{aligned}$$

The diastolic blood pressure before injection is given to three significant figures. So we also round the slope to three significant figures: 0.880.

The diastolic blood pressure after injection is given to the nearest whole number, so we round the intercept to one decimal place: 4.2.

So, the regression line is $y = 4.2 + 0.880x$.

To find the coordinates of two well-separated points on the line, we choose two well-separated values of x on the scatterplot, say $x = 100$ and $x = 130$.

When $x = 100$,

$$y = 4.2 + 0.880 \times 100 = 92.2,$$

so one point on the line is (100, 92.2).

When $x = 130$,

$$y = 4.2 + 0.880 \times 130 = 118.6,$$

so a second point on the line is (130, 118.6).

The scatterplot with the regression line is shown in the following figure.

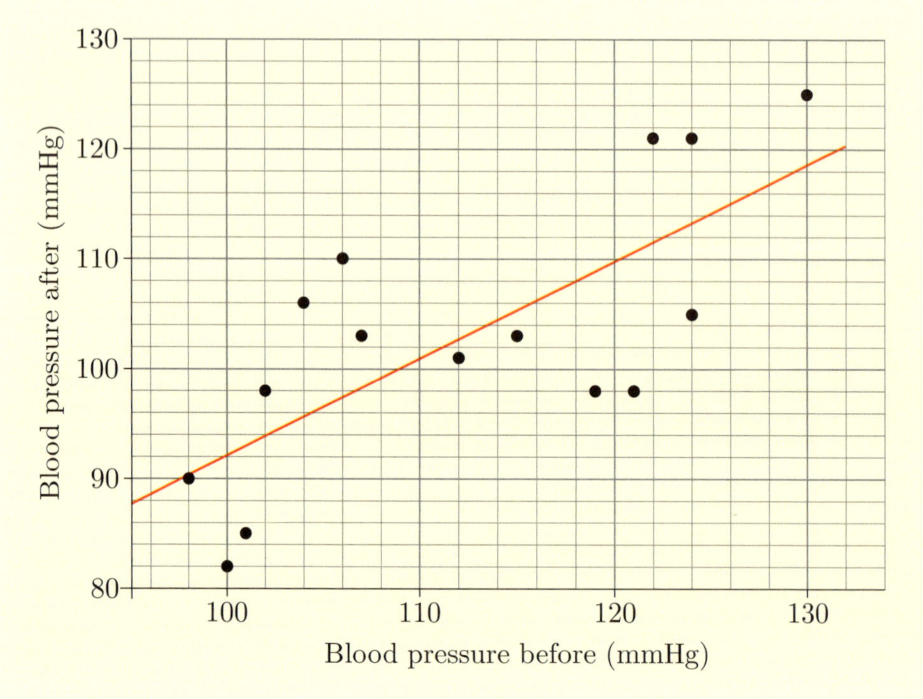

Scatterplot of blood pressure data from captopril study, with least squares regression line

Solution to Activity 18

(a) Yes, a straight-line model is reasonable for these data. There is no obvious pattern in the residual plot. The points are evenly scattered around the line $y = 0$.

(b) There is an obvious pattern in the residual plot. The residuals tend to be negative for small and large values of x, and positive for moderate values of x. So a straight-line model is not reasonable for these data.

(c) There is no obvious pattern in this residual plot, so a straight-line model is reasonable for these data. However, a couple of residuals stand out in the residual plot, one particularly big and the other particularly small. This suggests that there are a couple of outliers in the data, that is, a couple of points that do not fit the straight-line model as well as the rest of the data.

Solution to Activity 19

(a) For the first data point, (130, 125):

$$\begin{aligned}\text{Fit} &= 4.2 + 0.880 \times 130\\ &= 4.2 + 114.4 \simeq 119.\end{aligned}$$

So,

$$\text{Residual} \simeq 125 - 119 = +6.$$

The following table shows the fit and residual values (rounded to the nearest whole number) for the five data points.

x	y	Fit	Residual
130	125	119	+ 6
122	121	112	+ 9
124	121	113	+ 8
104	106	96	+10
112	101	103	− 2

(b) There is no obvious pattern to be seen here, which suggests that the least squares regression line is a reasonable model for the data.

Solution to Activity 20

(a) The regression line is $y = 4.2 + 0.880x$. So the predicted blood pressure two hours after injection of captorpril for a patient with an initial blood pressure of 105 mmHg is

$$4.2\,\text{mmHg} + 0.880 \times 105\,\text{mmHg} \simeq 97\,\text{mmHg}.$$

(b) For the regression line used in Example 19, x is the percentage of men unemployed in a town, and y is the percentage of households with no car. (This regression line was calculated in Examples 15 to 17.) Thus, for a town with 3.78% of men unemployed in 2001, the expected percentage of households with no car in 2001 is

$$5.27 + 4.42 \times 3.78 \simeq 22.0.$$

Solution to Activity 21

(a) For patients whose initial blood pressure is 110 mmHg, their blood pressure two hours after injection with captopril will on average be 101 mmHg.

(b) Suppose we looked at towns where the male unemployment was 4.00% in 2001, and in each of those towns we noted the percentage of households with no car. Then the average of those percentages would be close to 23.0%.

Solution to Activity 22

(a) This is reasonable. The data seem to relate to the same type of students (15-year-olds) that were used to fit the line. Also the value $x = 400$ is within the range of the x-values in Figure 51.

(b) This would not be reasonable. The relationship between the performance of an average 12-year-old on the reading scale and the performance of an average 15-year-old on the mathematics scale is not likely to be the same as that shown in Figure 51 (which is comparing average performances for 15-year-olds).

(c) This is not likely to reasonable. There is no guarantee that the straight-line relationship shown in Figure 51 is still going to apply for x-values as low as 200.

(d) This probably would be reasonable. The value $x = 575$ is a bit higher than the x-values plotted in Figure 51, but not by much. So the straight line will probably still be valid.

Solutions to exercises

Solution to Exercise 1

(a) As the weights are measured for different children from those whose heights are measured, the two variables height and weight are not linked.

(b) Even though the children are in the same school, the children whose weight is measured are still different from the children whose height is measured. So height and weight are still not linked.

(c) In this situation, heights and weights are measured for the same children. So height and weight are linked here.

Solution to Exercise 2

(a) The response variable is the average house price, and the explanatory variable is the calendar year, because it makes sense to think of variation in house price being explained by the calendar year but not the other way round.

(b) Here either variable could be regarded as the response variable, making the other variable the explanatory variable. This is because variation in men's wages and variation in women's wages may be related, but they probably vary together, rather than a change in one causing a change in the other.

(c) The response variable is employment rate, as this is the quantity to be predicted. The other variable, unemployment rate, is therefore the explanatory variable.

Solution to Exercise 3

There appears to be a positive linear relationship between a man's average hourly wage in a sector of the UK economy and the corresponding average hourly wage for a woman. This relationship appears to be reasonably strong. All of the sectors seem to fit with this general relationship, none standing out as particularly unusual.

Solution to Exercise 4

In this scatterplot, the relationship between house price and year appears to be positive and non-linear. So house prices generally went up during the period, but not always at the same rate. (House prices appear to have increased most quickly between about 2001 and 2004.) The relationship appears to be very strong. Arguably, the house prices in 2007 do not follow the same pattern as all the other years. The average house price in 2007 appears to be an outlier as it is higher than the relationship suggests it should have been.

Solution to Exercise 5

One suitable line is shown below. Your line should be similar to this, though it is not expected that it will match it exactly.

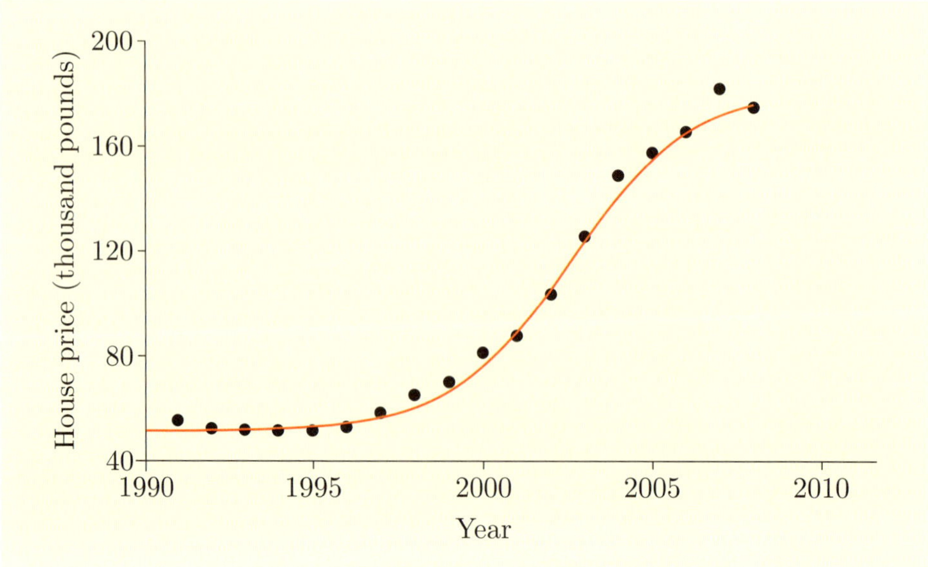

Notice that the line is a smooth curve, not a straight line. This fits with the relationship being non-linear. It does not go through all of the points. In particular it does not go through the point representing house prices in 2007, a point that was identified as a possible outlier in the solution to Exercise 4.

Solution to Exercise 6

(a) Fit $= 125 - 6 \times 20 = 125 - 120 = 5$.

(b) Fit $= -3 + 0.25 \times 20 = -3 + 5 = 2$.

(c) Fit $= 0.15 \times 20 = 3$. So Residual $= 4 - 3 = 1$.

(d) Fit $= 8 + 20 = 28$. So Residual $= 4 - 28 = -24$.

Solution to Exercise 7

(a) There appear to be slightly more negative residuals than positive residuals. So the fit of the line could be improved by moving the line down slightly. However, this imbalance is slight, so it is reasonable to conclude that the line fits the data well enough not to need adjusting.

(b) There are lots of positive residuals and few negative residuals. Furthermore, the negative residuals are a lot closer to the line corresponding to a residual of zero than the positive residuals are. So the line does not fit the data very well, and needs to be moved higher.

(c) The residuals tend to be positive for small values of the explanatory variable, and negative for large values. So the line does not fit the data very well, and needs to be made less steep.

(d) The residuals are generally evenly balanced, positive and negative, with the positive residuals spread through the range of the explanatory variable. So the line fits the data well.

Solution to Exercise 8

Line B cannot be the least squares regression line because it does not go through the point $(\overline{x}, \overline{y})$. Line A tends to be further away from the points than line C. So the sum of squared residuals is larger for line A than for line C. So out of the three, line C must be the least squares regression line.

Solution to Exercise 9

1. The four initial sums required are as follows.

$$\sum x = 19\,845, \quad \sum y = 355.6, \quad \sum x^2 = 39\,382\,485, \quad \sum xy = 706\,011.9.$$

2. The mean of the x-values and the mean of the y-values are

$$\overline{x} = 19\,845/10 = 1984.5$$

and

$$\overline{y} = 355.6/10 = 35.56.$$

3. The sum of the squared deviations of the x-values is

$$\begin{aligned}\sum(x-\overline{x})^2 &= 39\,382\,485 - \frac{(19\,845)^2}{10}\\ &= 39\,382\,485 - 39\,382\,402.5\\ &= 82.5,\end{aligned}$$

and the sum of the products of the deviations of the x- and y-values is

$$\begin{aligned}\sum(x-\overline{x})(y-\overline{y}) &= 706\,011.9 - \frac{19\,845 \times 355.6}{10}\\ &= 706\,011.9 - 705\,688.2\\ &= 323.7.\end{aligned}$$

4. We can now calculate the slope, b, of the regression line:

$$\begin{aligned}b = \frac{\sum(x-\overline{x})(y-\overline{y})}{\sum(x-\overline{x})^2} &\simeq \frac{323.7}{82.5}\\ &\simeq 3.923\,636\,364 \simeq 3.924.\end{aligned}$$

(The year is given to four significant figures, so the gradient is also rounded to four significant figures.)

5. The intercept, a, is then:

$$\begin{aligned}a &= \overline{y} - b \times \overline{x}\\ &\simeq 35.56 - (3.923\,636\,364 \times 1984.5)\\ &\simeq 35.56 - 7786.456\,364 = -7750.896\,364 \simeq -7750.90.\end{aligned}$$

(House prices are given to one decimal place, so the intercept is given to two decimal places.)

So, the regression line is $y = -7750.90 + 3.924x$.

Solution to Exercise 10

The correct matchings are

Scatterplot A and Residual Plot D; Scatterplot B and Residual Plot C; Scatterplot C and Residual Plot A; Scatterplot D and Residual Plot B.

The least squares regression line looks like a reasonable summary for the data in Scatterplot A, because the residuals look to be evenly scattered around zero, with no obvious pattern.

The least squares regression line also looks like a reasonable summary for the data in Scatterplot C, because again the residuals appear evenly scattered around zero. However, there are a couple of potential outliers, corresponding to the point whose residual is more than +4, and the point whose residual is less than −4.

The least squares regression line is not a reasonable summary for the data in Scatterplot B. Here this line does not capture the non-linear nature of the relationship.

The least squares regression line is also not a reasonable summary for the data in Scatterplot D. The relationship in this scatterplot also appears to be non-linear, and in a more complicated way than in Scatterplot B.

Solution to Exercise 11

(a) For 2009, the fitted value is

$$-16\,992.0 + 8.546 \times 2009 = -16\,992.0 + 17\,168.914 = 176.914.$$

So the predicted average house price is £177 000 (rounded to the nearest £1000).

For 2010, the fitted value is

$$-16\,992.0 + 8.546 \times 2010 = -16\,992.0 + 17\,177.46 = 185.46.$$

So the predicted average house price is £185 000 (rounded to the nearest £1000).

For 2030, the fitted value is

$$-16\,992.0 + 8.546 \times 2030 = -16\,992.0 + 17\,348.38 = 356.38.$$

So the predicted average house price is £356 000 (rounded to the nearest £1000).

(b) The years 2009 and 2010 are only slightly beyond the range of the data, so it is not that unreasonable to make predictions for these years. However, 2030 is far beyond of the range of the data, making the prediction of house prices in 2030 unreliable. Even if the model fits perfectly between 1990 and 2008, many things may happen between 2008 and 2030 to make it inappropriate by 2030. (As it turns out, the financial turmoil in the UK economy starting in the latter half of 2008 means that the model is of questionable use for predicting house prices in 2009 and 2010.)

(c) There is a distinct pattern in the residual plot. The residuals are negative during the period 1995 to 2003 and positive elsewhere. This suggests that the least squares regression line does not adequately model the data. A curved line is needed instead. This means even the predictions for average UK house prices in 2009 and 2010 now seem dubious.

Acknowledgements

Grateful acknowledgement is made to the following sources:

Table 1 HMSO (2004) *Census 2001: Key Statistics for Local Authorities in England and Wales*, Table KS09b. Crown copyright material is reproduced under Class Licence Number C01W0000065 with the permission of the Controller, Office of Public Sector Information (OPSI)

Table 2 HMSO (2004) *Census 2001: Key Statistics for Local Authorities in England and Wales*, Table KS17. Crown copyright material is reproduced under Class Licence Number C01W0000065 with the permission of the Controller, Office of Public Sector Information (OPSI)

Table 3 HMSO (2004) *Census 2001: Key Statistics for Local Authorities in England and Wales*, Tables KS09b and KS17. Crown copyright material is reproduced under Class Licence Number C01W0000065 with the permission of the Controller, Office of Public Sector Information (OPSI)

Figure 6 Bennett, G.W. (1988) 'Determination of anaerobic threshold', *Canadian Journal of Statistics*, John Wiley & Son Limited

Figure 7 Taken from: www.scotland.gov.uk/Publications/2010/11/05111814/42

Figure 11 Castillo, E., Hadi, A.S. and Minguez, R. (2009) 'Diagnostics for non-linear regression', *Journal of Statistical Computation and Simulation*, Taylor Francis Journals

Figure 14 HMSO (2011) *Social Trends 41 – Health*. Crown copyright material is reproduced under Class Licence Number C01W0000065 with the permission of the Controller, Office of Public Sector Information (OPSI)

Figure 15 OECD

Figure 17 HMSO (2003) *Census 2001: Key Statistics for Local Authorities in England and Wales*, Table KS11. Crown copyright material is reproduced under Class Licence Number C01W0000065 with the permission of the Controller, Office of Public Sector Information (OPSI)

Figure 18 Taken from: www.ons.gov.uk/ons/publications/index.html. Crown copyright material is reproduced under Class Licence Number C01W0000065 with the permission of the Controller, Office of Public Sector Information (OPSI)

Figure 21 Taken from: www.ons.gov.uk/ons/rel/lms/labour-market-statistics/september-2011/earnings-tsd-dataset.html. Crown copyright material is reproduced under Class Licence Number C01W0000065 with the permission of the Controller, Office of Public Sector Information (OPSI)

Figure 22 Taken from: www.nationwide.co.uk/hpi/datadownload/data_download.htm

Figure 34 MacGregor, G.A., et al. (1979) 'Blood pressure from captopril study', *British Medical Journal*

Subsection 1.3 cartoon, www.causeweb.org

Subsection 1.4 photo of a sphygmomanometer: Joey Parsons/Flickr.com

Subsection 1.4 household expenses: Crabchick/Flickr.com

Subsection 1.4 photo of tomatoes: Andrew Frogg/Flickr.com

Subsection 1.4 depiction of children: George Rex/Flickr.com

Subsection 2.1 figure, 'Lots more positive relationships', taken from: http://organizationalpositivity.com/?p-48

Subsection 2.1 figure, equipment to measure oxygen uptake: Active Steve/Flickr.com

Subsection 2.4 cartoon ('your theory is wrong'), taken from: www.thewired.be/?p=293

Subsection 3.1 logo, www.fitline.com

Section 5 cartoon ('data don't make any sense'), VALDO

Subsection 5.2 cartoon (lottery statistics), www.causeweb.org

Subsection 5.3 figure ('my hobby: extrapolating'): www.xkcd.com. This file is licensed under the Creative Commons Attribution-Non commercial Licence http://creativecommons.org/licenses/by-nc/3.0/

Photo in solution to Activity 17: Courtesy of the International Slide Rule Museum

Every effort has been made to contact copyright holders. If any have been inadvertently overlooked the publishers will be pleased to make the necessary arrangements at the first opportunity.

Index